D0946506

# FUNDAMENTALS OF PREPARATIVE ORGANIC CHEMISTRY

**IMPORTANT ERRATA**

p.118 : Acid chlorides ............
Add **to** a large excess of water ...

Dimethyl sulphate
Add **to** dilute NaOH or $NH_4OH$...

# FUNDAMENTALS OF PREPARATIVE ORGANIC CHEMISTRY

R. KEESE, Dipl.Chem., Dr.rer.nat.
Professor, Institute of Organic Chemistry
University of Berne, Switzerland

R. K. MÜLLER, Dipl.Chem. ETH, Dr.sc.techn.
Department of Vitamin and Nutritional Research
F. Hoffmann-La Roche & Co. Ltd.
Basle, Switzerland

and

T. P. TOUBE, B.Sc.(Hons.), M.Sc., Ph.D.
Department of Chemistry
Queen Mary College, University of London

Illustrated by H. Brühwiler

*Developed and compiled in cooperation with*
Hans-Ulrich Blaser, Henri Martin Dubas, Sallem Farooq, Walter Fuhrer, Peter Gygax, Franz Heinzer, Reinhard Hobi, Ernst-Peter Krebs, Peter Michael Müller, Andreas Pfaltz, Armin Pfenninger, Erich Stamm

**ELLIS HORWOOD LIMITED**
Publishers · Chichester

Halsted Press: a division of
**JOHN WILEY & SONS**
New York · Brisbane · Chichester · Toronto

ALBRIGHT COLLEGE LIBRARY

First published in 1982 by

**ELLIS HORWOOD LIMITED**
Market Cross House, Cooper Street, Chichester, West Sussex, PO19 1EB, England

*The publisher's colophon is reproduced from James Gillison's drawing of the ancient Market Cross, Chichester.*

**Distributors:**

*Australia, New Zealand, South-east Asia:*
Jacaranda-Wiley Ltd., Jacaranda Press,
JOHN WILEY & SONS INC.,
G.P.O. Box 859, Brisbane, Queensland 40001, Australia

*Canada:*
JOHN WILEY & SONS CANADA LIMITED
22 Worcester Road, Rexdale, Ontario, Canada.

*Europe, Africa:*
JOHN WILEY & SONS LIMITED
Baffins Lane, Chichester, West Sussex, England.

*North and South America and the rest of the world:*
Halsted Press: a division of
JOHN WILEY & SONS
605 Third Avenue, New York, N.Y. 10016, U.S.A.

Translated from the German
*Grundoperationen der praparativen organischen Chemie: Eine Einfuhrung (3rd Edn.)* by R. K. Müller and R. Keese, Juris-Verlag, Zurich, 1981.

**©1982 R. Keese, R. K. Müller, T. P. Toube/Ellis Horwood Limited**

**British Library Cataloguing in Publication Data**
Keese, R.
Fundamentals of preparative organic chemistry. –
(Ellis Horwood Series in Organic Chemistry)
1. Chemistry, Organic – Experiments
I. Title II. Müller, R. K. III. Toube, T. P.
IV. Grundoperationen der präparativen organischen Chemie. *English*
547'0028 QD261

**Library of Congress Card No.** 82-3045 AACR2

ISBN 0-85312-396-9 (Ellis Horwood Limited – Library Edn.)
ISBN 0-85312-450-7 (Ellis Horwood Limited – Student Edn.)
ISBN 0-470-27522-7 (Halsted Press)

Typeset in Press Roman by Ellis Horwood Limited.
Printed in Great Britain by Unwin Brothers Ltd., of Woking.

**COPYRIGHT NOTICE –**
All Rights Reserved. No part of this publication may be reproduced, stored in a retrieval system, or transmitted, in any form or by any means, electronic, mechanical, photocopying, recording or otherwise, without the permission of Ellis Horwood Limited, Market Cross House, Cooper Street, Chichester, West Sussex, England.

547.076
K26f

184606

# Table of Contents

# Introduction

## INTRODUCTION

The transmission to their students of 'know-how' in the field of experimental organic chemistry is one of the chief problems facing Teaching Assistants in the chemistry laboratory. Assistants, postgraduate student Demonstrators, and undergraduate students come and go; the preservation and development of experimental skill and knowledge in an institution is thus essential. The publication of this introduction to the fundamentals of preparative organic chemistry is certainly to be welcomed from this point of view; I cannot doubt that its appearance will be welcomed by students. The joint authorship and the style of work allow something of the particular atmosphere that reigns in the laboratory under our Assistants to be sensed. Recognition for the creation and maintenance of this atmosphere is due in particular to the Chief Assistants, Dr Rolf Scheffold (1964–68), Dr Reinhart Keese (1968–74), and Dr Robert Karl Müller (since 1974). This publication is the result of their initiative, and I wish their book an effective and wide distribution.

A. Eschenmoser.
ETH,Zurich.
1975

# Preface

## Preface to the Orginal German Edition

Basic laboratory technique in organic chemistry has an essential place in the training of a chemist. It provides the foundation for preparative experimental skills and lays the groundwork for subsequent independent research. Since the successful synthesis of a substance often depends on the work-up of the reaction mixture, separation techniques are particularly important. A systematic approach to work increases one's pleasure in experimentation, so that organisational aspects, e.g. keeping of laboratory notebooks, or preparation of reports, also need to be mastered.

It is our opinion that the one-semester course in basic organic chemical laboratory practice which has for some years been in operation at the ETH, Zurich (and more recently at the University of Berne) has succeeded in inculcating the above concepts. In the first section the most important separatory techniques used in synthetic organic chemistry are covered initially. Associated with each type of separation is the resolution of a multi-component mixture by 'shaking out', the components being chosen to be susceptible to complete identification by instrumental methods. Then follows an approximately equal period devoted to preparative chemistry, introduced by the material we have labelled 'Hints on the synthesis of organic compounds'.

One section of this manual, which has been developed over the last few years in particular by the assistants, has previously been produced for our basic laboratory course as a collection of loose sheets. In our opinion, however, these 'Fundamentals' play such a vital part in the education of an organic chemist that we have decided to make a revised version of the material more generally available. As each section has a strong practical bias this book is suitable for use in courses in synthetic organic chemistry as well as for those subjects where chemistry plays a minor part. Our experience has shown that practical hints are more valued in the laboratory than extensive discussions of theoretical principles, and we believe our approach will make this manual of use wherever synthetic organic chemistry is undertaken.

We thank Dr J. Schreiber and Administrator R. Kempf for permission to use the section on 'Waste Disposal and the Destruction of Dangerous Materials' prepared at the ETH, Zurich. In particular, we are grateful to Dr Schreiber for his professional advice over many years and for furnishing us with a number of references. For the critical appraisal of the manuscript we are indebted to Professor O. Schindler, Berne.

While the authors take full responsibility for the contents of this volume, they welcome any criticisms, corrections, and suggestions for improvement.

Zurich
September 1975

R. K. Müller
R. Keese

**Preface to the Third German Edition**

The continued use of *Grundoperationen* in the organic teaching laboratories at Zurich and Berne makes it necessary for the content to be matched to changing needs. The original version has therefore been revised and extended by the addition of new chapters and sections. The content now exceeds the scope of a foundation in practical chemistry. In our opinion, work under inert atmosphere and handling of isotopes are part of general knowledge in laboratory practice.

The revision owes a great deal to colleagues, co-workers, and specialists, and we wish to thank them.

Dr T. P. Toube, Queen Mary College, London, who has prepared the English translation of *Grundoperationen,* has made a number of suggestions for the expansion of the content. We thank Dr D. G. Buckley and Mr G. S. Coumbarides, Queen Mary College, London, for the first draft of the section on isotopes.

For the evaluation of sections of the manuscript we thank in particular Professor H. Gerlach and Professor M. Neuenschwander, PD Dr U. Krähenbühl, Dr H. Wetter, and Mrs V. Meyer, of the Institute for Organic Chemistry, Berne, Dr F. Berdat, Lonza AG, Visp, and Dr D. Bauer, Hoffmann-La Roche & Co. AG, Basle. We owe particular thanks to Dr J. Schreiber, who has once again given us the benefit of his considerable experience.

Berne, June 1981

Robert K. Müller
Reinhart Keese

CHAPTER 1

# Accident Prevention and First Aid

## 1. EYE INJURIES

Always wear safety glasses in the laboratory!

If any chemical gets into the eyes, the eye should be bathed at once with water at the nearest tap or eyebath. Washing should continue for at least ten minutes. The help of a second person in holding the head and spreading the eyelids to ensure thorough washing is desirable. This procedure is appropriate for accidents involving acids, strong bases (especially dangerous as they disintegrate the tissues and allow the contaminant to penetrate more deeply), and other chemicals.

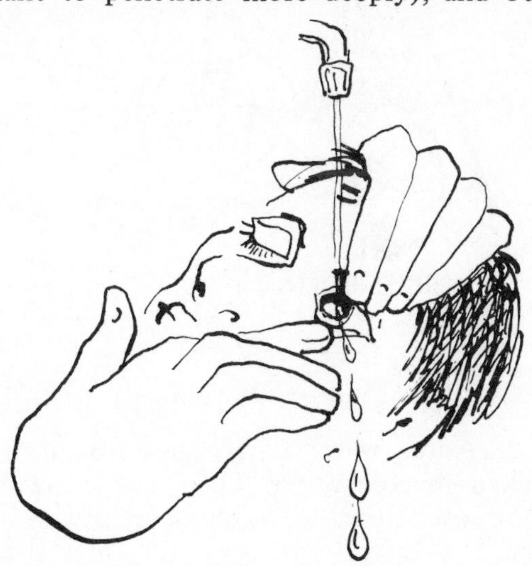

Under no circumstances should acid splashed in the eye be washed out with bases (or vice versa), as this generally does more harm than good. After the eyes have been thoroughly rinsed with water for a sufficient period, professional medical attention should be sought. If contact lenses are worn, they should be removed immediately: chemicals may be drawn under the lenses by capillary action and cannot easily be rinsed out.

## 2. FIRE AND BURNS

Know where fire extinguishers, fire blankets, showers, and fire alarm buttons are situated in the laboratory, and find out how they are used. Read the fire regulations and make sure you understand them.

Never smoke in the laboratory. If you have to smoke, go out to a safe area. Never discard cigarettes or matches without extinguishing them under a tap first. Smouldering cigarette ends thrown into drains regularly cause fires!

Open flames should only be used where other sources of heat cannot be employed or are inappropriate. Bunsen burners should be used in the fume hood. Ensure that there are no flammable materials in the vicinity before lighting the flame. Many fires are caused by ineffective destruction of reactive reagents (see pp 118–9).

— FORBIDDEN —

### Skin burns

Cool at once with running water. Larger burns may need to be covered with cloths soaked in cold water. Then seek professional medical attention. Do not cover burns with oily or greasy ointments. Do not puncture blisters.

## 3. CUTS

Most cuts occur when working with glass. Many accidents can be prevented if glass is handled with a glass cloth or leather gloves, especially when being pushed through holes in stoppers, etc.

Take particular care to cool ampoules thoroughly in ice before opening them.

**Treatment**

– small cuts: allow to bleed, disinfect, bandage.

– larger cuts: if necessary stop bleeding (see below), cover with bandage, get medical attention.

– cut fingers: remove rings; unless they are obviously trivial, cuts on the fingers should be examined by a doctor.

**Bleeding**

(a) to stop bleeding:

- raise the affected limb
- if this is not sufficient, apply pressure at the appropriate pressure point (requires practice!) – or apply pressure directly over the wound, using a large dressing.

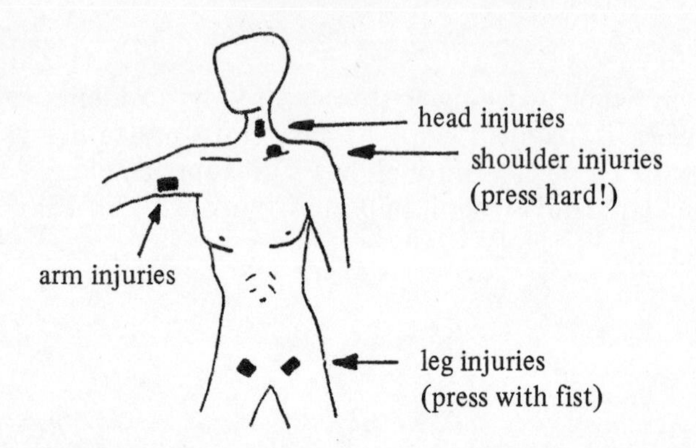


(b) to prevent resumption of bleeding, apply one or more pressure bandages.
(c) seek medical attention.

## 4. POISONING

Medical treatment is facilitated if the nature of the poison can be established (obtain a sample of the substance, gas, vomit, etc.).

**General Procedure**

(a) elimination of the poison.
(b) medical treatment.

(i) *Oral poisoning*

Induce vomiting by giving warm salt water (3 heaped teaspoons NaCl per glass) to drink. Repeat until vomit is clear. Then seek medical aid.

Do **not** induce vomiting if unconscious.

Do **not** induce vomiting if solvent has been swallowed.

If solvent has been swallowed do not induce vomiting, but give 200 $cm^3$ pure liquid paraffin to drink. Then seek medical aid. If acid has been swallowed give plenty of water to drink, followed by milk of magnesia; if alkali, give plenty of water to drink, followed by vinegar or 1% acetic acid.

(ii) *Gas poisoning*

Remove victim from the danger zone (wear compressed air breathing apparatus), keep him quiet, and transport to doctor on a horizontal stretcher. If breathing has stopped, apply artificial respiration by a method other than mouth to mouth, e.g. Silvester.

Painful coughing may be assuaged somewhat by inhalation of alcohol vapour (using a cottonwool pad soaked in ethanol).

(iii) *Percutaneous poisoning*

Remove contaminated clothing immediately, wash affected area thoroughly, then seek medical assistance.

In acute cases, the *time factor is critical.* Carry out the above first aid procedures and get medical attention immediately. Clothing must be washed and aired before re-use.

Knowledge of the toxicity of chemical substances is an essential part of the training of a responsible chemist.

(iv) *Special hazards*

Concentrated sulphuric acid

When using concentrated sulphuric acid always wear safety spectacles and rubber gloves. Concentrated sulphuric acid spillages onto the skin should first be mopped up as rapidly as possible, using cottonwool or cloth before the area is rinsed with plenty of cold water. Then seek medical attention.

Hydrogen fluoride (hydrofluoric acid)

Particular care is needed in handling this substance. Contact with hydrogen fluoride (liquid or vapour) is exceptionally dangerous. Injuries are extremely painful and are difficult to treat. If you are going to use HF, please read up the published safety procedures and implement them.

(v) *Chemical Carcinogens*

The transformation of normal cells into cancer cells appears to be an irreversible process. In considering carcinogenesis, one needs to take into account every encounter with a carcinogen: even a single exposure can result in tumour formation. The latent period between the initial chemical insult and the appearance of malignant symptons can be very long, reaching as much as 40 years in Man. This means that – in contrast to acute poisons – no immediate or clearcut connection between cause and effect may be evident. *Extreme care is therefore essential when working with carcinogens.* The following advice should be followed:

- work with carcinogenic substances only if you cannot avoid it;
- less dangerous reagents or solvents can often be substituted for carcinogenic ones in chemical reactions;
- one should try to *choose synthetic routes* which avoid carcinogenic reagents and intermediates;
- the risk can be reduced by careful technique and the use of protective and safety equipment: work in a fume hood, wear dust-masks and disposable gloves, cover the bench with foil;
- work with carcinogens only in closed apparatus;

– destroy excess reagents immediately and, if appropriate, hand in reaction residues for safe disposal.

Some classes of compound in which carcinogens have been found:

Alkylating agents

$(MeO)_2SO_2$
dimethyl sulphate**

$CH_2N_2$
diazomethane**

NH

aziridine**

$CH_3I$
iodomethane**
(methyl iodide)

O
O

β-propiolactone**

$(ClCH_2)_2O$
dichloromethyl ether***

Hydrazine and azo and azoxy compounds

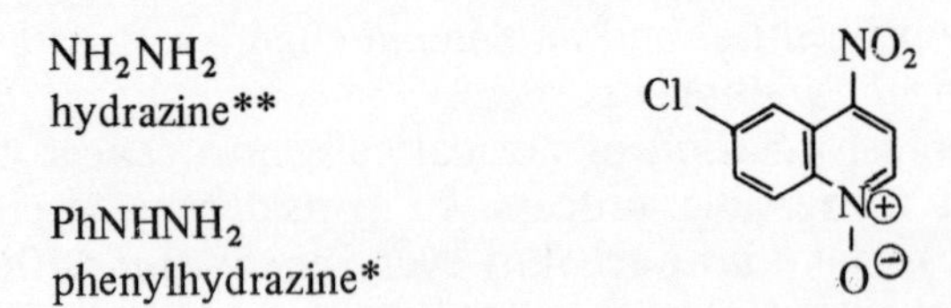

$NH_2NH_2$
hydrazine**

$PhNHNH_2$
phenylhydrazine*



6-chloro-4-nitroquinoline-1-oxide

$CH_3$ N—N=N—
$CH_3$

N,N-dimethyl-4-aminoazobenzene

Aromatic amines and azoxy and nitro compounds

$NH_2$ $H_2N$ $NH_2$

4-aminobiphenyl*** benzidine***

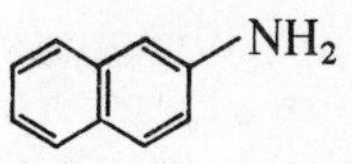

$NH_2$ (structure)

2-naphthylamine***

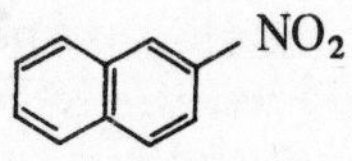

$NO_2$ (structure)

2-nitronaphthalene**

Aromatic hydrocarbons

benzene***

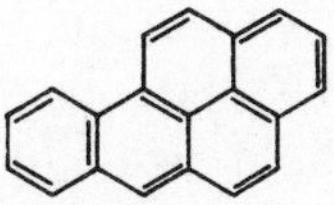

3,4-benzpyrene***

Nitrosamines and nitrosamides

$(CH_3)_2N$-NO
N,N-dimethylnitrosamine**

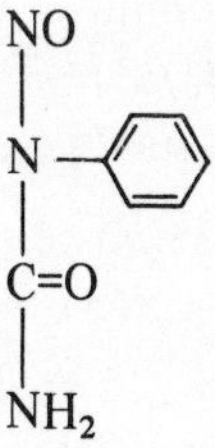

NO
|
N — (phenyl)
|
C=O
|
$NH_2$

N-nitroso-N-phenylurea

Natural products

*e.g.* aflatoxins, safrole, isosafrole, cycasine, pyrrolizidine alkaloids

Inorganic compounds

arsenic compounds***
nickel and its compounds***
beryllium and its compounds***
zinc chromate***
cadmium and its compounds*

asbestos***
chromates**
antimony trioxide*
chromium trioxide*
chromium carbonyl*

Miscellaneous

coal tar***
vinyl chloride***
acetamide

bitumen***
thiourea

**Footnote**

*** known to produce malignant tumours in Man
** unequivocally carcinogenic in animals
* suspected carcinogens on the most recent results
*No asterisk* indicates *no assignment of degree of danger possible* on the basis of present knowledge

It is certain that not all carcinogenic substances and reagents have yet been identified. One should therefore take care that one develops responsible and hygienic working habits.

**Bibliography**

*Handbook of Laboratory Safety,* N. V. Steere ed., Chemical Rubber Co., 2nd edition, 1971.

M. I. Sax, *Dangerous Properties of Industrial Materials,* Van Nostrand Reinhold, 3rd edition, 1968.

*Chemical Carcinogens,* C. E. Searle ed., ACS Monograph 173, American Chemical Society, 1976.

*Hazards in the Chemical Laboratory,* L. Bretherick ed., Royal Society of Chemistry, 3rd edition, 1981.

*Guide for Safety in the Chemical Laboratory,* 2nd edition, Manufacturing Chemists Association, Van Nostrand Reinhold, 1972.

# CHAPTER 2

# Crystallisation

Crystallisation is one of the most effective purification techniques for solids. Crystalline compounds are generally more stable and easier to handle than solutions or oils, and can be effectively characterised and identified. Furthermore, good crystals are a *sine qua non* for an X-ray structure analysis.

Crystals can be obtained from the melt (supercooled liquid phase), the vapour phase (sublimation), or from a supersaturated solution. It is crystallisation from supersaturated solution which is most commonly employed.

Before carrying out a crystallisation it is an advantage if one has some idea of how pure the material is and of the nature of the probable impurities. One can often use the normal methods for estimating purity (t.l.c., m.p., i.r., n.m.r., etc.) although a closer study of the reaction (starting materials, side products, etc.) may need to be carried out. The same criteria of purity enable one to test the success of the crystallisation. Samples should always be recrystallised to constant m.p. Products that stubbornly refuse to improve in purity may need sterner measures (chromatography, extraction, etc.).

**Workplan for a Crystallisation**

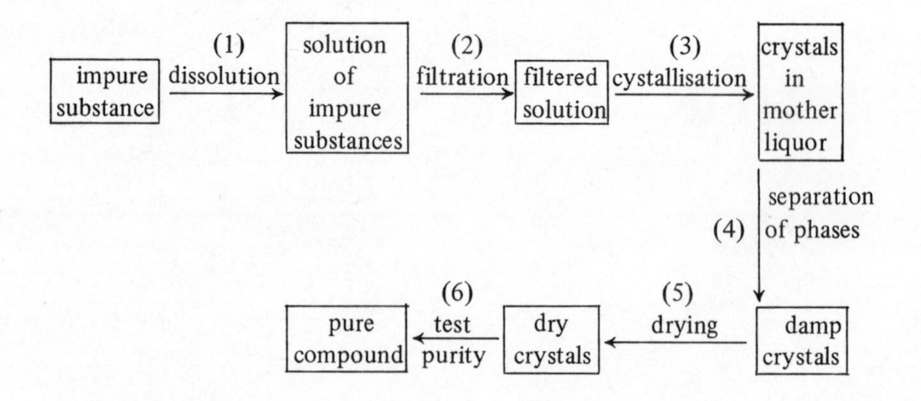


(1) *Dissolution*

Preliminary tests for solubility can be carried out in ignition tubes using small amounts of the material and solvent, first in the cold and then heated. For cystallisation one should always use redistilled solvents.

(a) Substances tend to be most soluble in chemically similar solvents.

| Class of substance | | | Efficient solvents |
|---|---|---|---|
| Hydrocarbons | hydrophobic (lipophilic) non-polar | | pentane, hexane, petroleum ether, benzene |
| Ethers | | | diethyl ether, methylene chloride |
| Halohydrocarbons | | increasing polarity ↓ | chloroform |
| Tertiary amines | | | acetone |
| Esters | | | |
| Ketones and Aldehydes | | | ethyl (or methyl) acetate |
| Phenols | | | |
| Alcohols | | | ethanol |
| Carboxylic acids | | | methanol |
| Sulphonic acids | | | water |
| Organic salts | hydrophilic polar | | |

(b) Good crystallisation media: very soluble in hot solvent, insoluble in cold.

(c) Poor solvents often produce poor crystals.

(d) Very good solvents require very high concentrations of solute for crystallisation.

(e) Polar solvents tend to produce better crystals than hydrocarbon solvents.

(f) Mixed solvents often prove best for crystallisation.

It is a good idea to try different solvents to find the one which gives the most efficient crystallisation (e.g. using the series of increasing polarity as a guide). If the substance is already crystalline don't use the whole sample for your recrystallisation; save a few milligrams for seeding (and for t.l.c. comparison, if appropriate). The purer the substance and the larger its crystals, the more slowly it will dissolve. Large crystals can be ground before adding solvent. Dissolution takes time, and in some cases the substance will have to be heated under reflux in the appropriate solvent in order to obtain a sufficiently concentrated solution. For thermally labile materials, mixed solvents are most appropriate (see (3) below).

Weigh the crude material before crystallising so that the yield of purer material can be determined. The differences between the masses of crude and pure material should equal the mass of the material obtained by evaporation of the mother liquor.

(2) *Filtration*

Filtration serves to remove dust and insoluble impurities. It is often sufficient to filter the hot solution through a funnel fitted with a cottonwool plug. To prevent premature crystallisation excess solvent can be used, and the solution concentrated to the correct volume afterwards. It is, however, better to filter through a sintered glass suction filter of the appropriate grade. If not available, filter through a fluted filter paper of the appropriate grade which has been pre-heated by pouring through it a portion of the hot solvent. Note that organic material is likely to be eluted from the filter paper.

If the solution is strongly coloured by impurities, activated charcoal decolorisation is necessary. The material is first dissolved in a polar solvent (e.g. acetone, ethyl acetate, ethanol) and then treated with 2–4% of its weight of charcoal for 10 minutes with stirring (and if necessary heating). The solution is then filtered under suction through a pad of Celite (a filtration aid: a suspension of Celite is formed into a pad on the filter under suction; it holds back even fine particles of charcoal and avoids clogging of the filter).

(3) *Crystallisation*

Supersaturated solutions needed for crystallisation can be prepared in the following ways:

(a) by slowly cooling a hot saturated solution to room temperature or below (using ice or refrigeration) (suitable for the majority of substances which are more soluble in hot solvent than in cold);

(b) by slow evaporation of solvent from an open vessel – undesirable, but useful in stubborn cases, and also often advisable for the preparation of samples for X-ray analysis;

(c) by slowly adding a miscible 'poor' solvent to a solution until it just starts to go cloudy (mixed solvent technique). Typical mixed solvents are dichloromethane-hexane, chloroform-hexane, ether-hexane, ether-acetone, acetone-water, methanol-water. If possible choose a system in which the better solvent is the one with the lower boiling point.

Crystallisation often refuses to occur spontaneously even when the solution is supersaturated, and must be induced by the formation of crystal nuclei.
The following techniques can be used:

(a) addition of a seed crystal (saved from (1), or in extreme cases even a homologous compound);

(b) scratching the side of the vessel with a glass rod;

(c) cooling in solid $CO_2$ and scratching with a glass rod as the solution slowly warms to room temperature.

Rules of thumb:
(i) Maximum formation of nuclei occurs *ca.* 100°C below the m.p.
(ii) Rapidity of crystallisation is greatest *ca.* 50°C below the m.p.
(iii) Vigorous cooling does not necessarily lead to more rapid crystallisation.

If the substance comes down as an oil, one can try crystallisation from more dilute solution or with less rapid cooling. If a compound completely refuses to crystallise it may first require further purification (e.g. chromatography).

(4) *Separation of Phases*
The mother liquor is removed from the crystals by filtration, preferably through a glass sinter, or under an inert gas (nitrogen, argon) in a special filtration apparatus. For small samples it may be possible to draw off the mother liquor using a finely drawn pipette.

The crystals should then be washed on the filter with solvent of the same composition as the mother liquor: if this is not done in the case of mixed solvents, the single solvent used will either dissolve the crystals or else precipitate impurities from the traces of mother liquor still clinging to the crystal surfaces!

(5) *Drying*
Crystals should be dried to constant weight in a desiccator or other suitable vessel, usually under reduced pressure (water pump, or high vacuum with cold trap). Thermally stable compounds can be warmed

to *ca.* 50°C below their m.p. as long as their volatility is not too great (excessive drying under vacuum can lead to loss of material by sublimation). Drying agents should only be used in desiccators when crystals from aqueous solvents are being dried.

### (6) *Tests for Purity*

The efficacity of a recrystallisation may be judged using one or more of the following criteria.

#### Melting point (m.p.)

The m.p. is the most common test for purity after recrystallisation. Even small amounts of impurity may depress the m.p. appreciably. In general, one recrystallises to constant m.p. A further increase in purity may be obtained by sequential recrystallisation – it may pay to crystallise from an alternative solvent system. Material should be carefully dried and finely pulverised for m.p. determinations. An estimate of the m.p. range should first be made, heating the sample comparatively rapidly. A second sample is then heated more slowly (*ca.* 2°C per minute) from about 20°C below the estimated value to obtain the exact m.p. Melting points are usually determined in capillary tubes (m.p. tubes) open to the air. In special circumstances sealed evacuated capillaries may be necessary. A heated stage microscope (Kofler block) is sometimes used especially when the m.p. of a very small sample is required.

The result may be quoted thus: m.p. 109–110°C (uncorrected, open capillary), lit. [ref.] m.p. 110°C.

#### Mixed m.p.

A mixed m.p. serves to identify a substance. An intimate 1:1 mixture of the unknown compound and of a pure sample of the reference compound is prepared. Three m.p. tubes are used containing the unknown, the reference, and the mixture respectively, and they are heated at the same time in the same apparatus.

(A) If all three melt simultaneously the unknown is identical to the reference compound.

(B) If the mixture has a m.p. between those of the other two (and the unknown has the lowest m.p.), the unknown is probably an impure sample of the reference compounds.

(C) If the mixture has the lowest m.p., the other two substances are different compounds (even if they fortuitously have identical m.p.).

Thin Layer Chromatography (t.l.c.)
See under *Chromatography* (p. 36).

Spectroscopy
See in relevant texts on IR, UV, NMR, MS, etc.

**Fractional Crystallisation**
This is a technique for separating two substances by repeated crystallisation (e.g. separation of racemates via their crystalline diastereoisomeric derivatives).

A schematic representation of the process is shown below (C = crystals, ML = Mother-liquor).

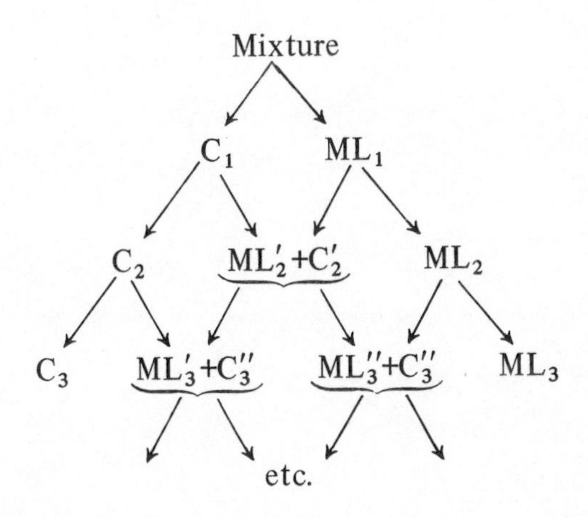


Pure crystals of the less soluble component | Mixtures | Samples enriched in the more soluble component

The advent of effective spectroscopic methods for the identification of organic compounds has decreased the traditional use of crystalline derivatives. However, crystalline derivatives are still a useful aid for identification. A list of such derivatives for the various

184606
ALBRIGHT COLLEGE LIBRARY

classes of compounds can be found in e.g. R. L. Shriner, R. C. Fuson, D. Y. Curtin, and T. C. Morrill, *Systematic Identification of Organic Compounds,* Wiley (1964).

CHAPTER 3

# Distillation

Distillation is one of the chief techniques for separating mixtures of liquids. Separation relies on the difference in vapour pressure of the various components of the mixture. The mixture is vaporised by heating and the vapour is then condensed, and in the process the vapour (and therefore the condensate) becomes enriched in the more volatile components.

**Some theorectical considerations**

*The Clausius-Clapeyron Relationship*

The vapour pressure ($P$) of a liquid rises rapidly with temperature ($T$). If the molar heat of vaporisation is $L_V$, the Clausius-Clapeyron Relationship gives

$$\frac{d \ln P}{dT} = \frac{L_V}{RT^2} \tag{3.1}$$

A graphical representation of an integrated from of (3.1) shows a straight line the slope of which is determined by the molar heat of vaporisation (which does not differ greatly between chemically similar materials). If one knows the boiling point (b.p.) of a liquid at a particular pressure, the b.p. at any other pressure can be estimated ($L_V$ is taken to be temperature independent in this treatment).

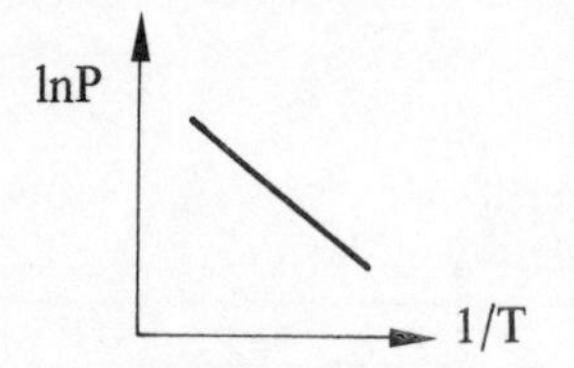


The slope is determined by the molar heat of vaporisation which is fairly constant between chemically similar substances.

Rules of thumb

(1) Halving the pressure reduces the b.p. by about 15°C.
(2) Waterpump vacuum (10–15 torr)† usually reduces the b.p. by about 100°C.
(3) Vacuum pump pressure (*ca.* 0.1 torr) causes a further reduction of about 60°C.

*Raoult's Law*

(i) Simple distillation

The enrichment of the more volatile component in the vapour of a binary mixture after a *single* evaporation is given by Raoult's Law as:

† 1 torr = 1.33322 mb.

$$\frac{y}{1-y} = \frac{P_a}{P_b} \cdot \frac{x}{1-x} \tag{3.2}$$

where $y$ = proportion of more volatile component in vapour
$x$ = proportion of more volatile component in pot
$P_a, P_b$ = vapour pressures by pure components a and b.

PRACTICAL EXAMPLE

With a 1:1 binary mixture (i.e. $x = 0.5$) having a b.p. difference of 60°C (giving $P_a/P_b \approx 10$), solution of equation (3.2) gives $y = 0.9$ (i.e. in the vapour there is a ratio of 9:1 in favour of the more volatile component, a).

IMPORTANT RULE OF THUMB

For simple distillation, tantamount to a single vaporisation process, really to be effective the components should differ in their boiling points by at least 80°C.

In practice this means that simple distillation should only be used to separate an already fairly pure volatile substance from high-boiling impurities (e.g. solvent from inorganic impurities or drying agents, reaction products from polymeric by-products, etc.). Equally, it can be used to remove all solvent from certain liquid reaction products.

(ii) Fractional Distillation (Rectification)

If the b.p. difference is too small for simple distillation to be effective, it is necessary to resort to repeated distillations. In practice one employs a **fractionating column** in which the vapour and condensed phases move in opposite directions. The efficiency of such columns is expressed in **theoretical plates,** a theoretical plate being defined as the column unit having the same effective separation as a simple distillation (and often expressed in cm of column height).

For an $n$-fold repetition of the vaporisation-condensation process, the enrichment of the more volatile component is given by

$$\frac{y}{1-y} = \left(\frac{P_a}{P_b}\right)^n \cdot \frac{x}{1-x} \tag{3.3}$$

where $x$, $y$, $P$ have the same meanings as in equation (3.2).

PRACTICAL EXAMPLE

With a 1:1 binary mixture (i.e. $x = 0.5$) having a b.p. difference of 30°C (i.e. $P_a/P_b \approx 3$), solution of equation (3.3) to give $y = 0.95$ requires in $n \approx 3$, i.e. one would need a column of at least 3 theoretical plates to get the more volatile component at least 95% pure.

## Distillation in Practice

*Choice of temperature and pressure*

An unknown thermally stable mixture should first be simply distilled at atmospheric pressure. Heating should not be continued above about 180°C even if material that will not distil remains in the flask. The distillation flask is cooled, the collected fractions removed and the distillation continued under reduced pressure. Higher boiling material should be distilled using a vacuum pump at *ca.* 0.1 torr. Heat-sensitive, unstable compounds should, of course, be distilled only under reduced pressure.

*Apparatus*

The choice of apparatus depends fundamentally on the nature of the problem, as shown by b.p. differences, types of impurity, amount of material, etc. For some suggestions, see the illustrations.

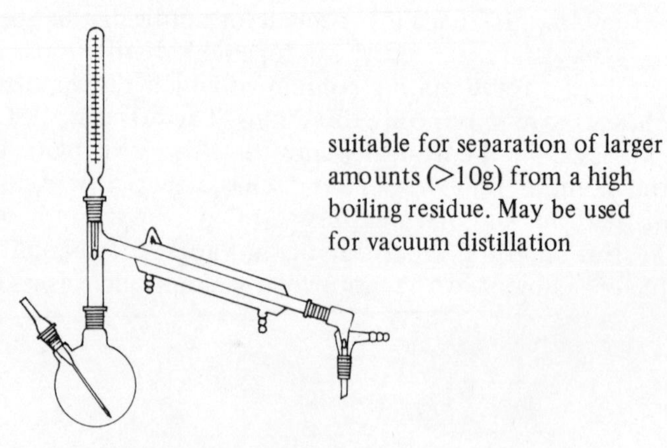

round bottomed flask with Claisen head and receiver bend

suitable for separation of larger amounts (>10g) from a high boiling residue. May be used for vacuum distillation

Apparatus for fractionation

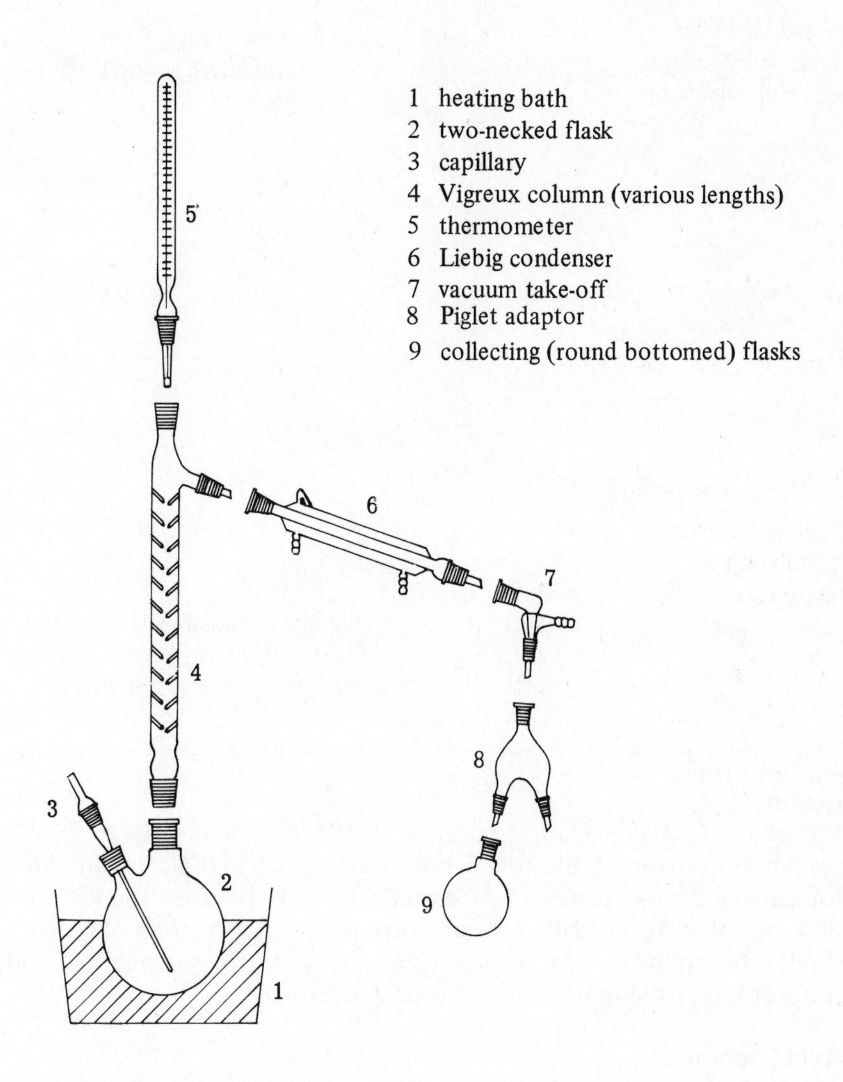


The figure shows the standard apparatus for fractional distillation. More effective separation requires the use of other columns (spinning band, packed, etc.) and a head allowing the reflux ratio to be adjusted.

round bottomed flask with K adapter and coil condenser

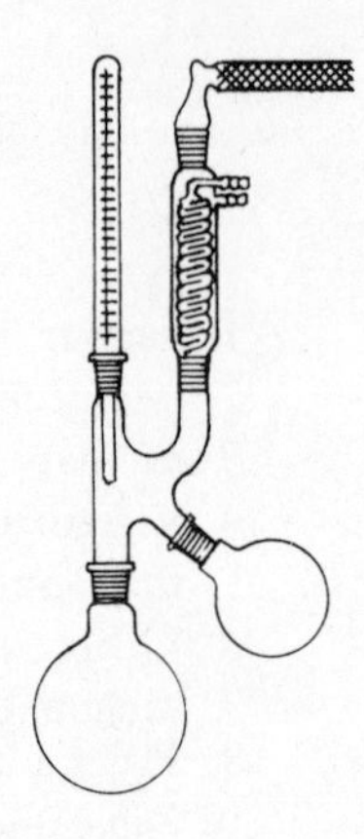

suitable for solvent distillation at atmospheric pressure

bulb (short path) distillation apparatus

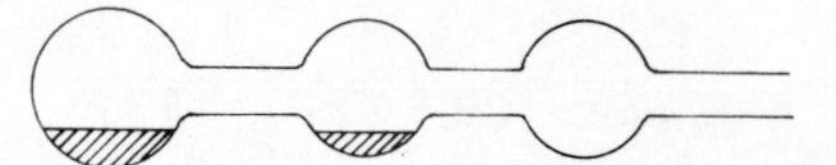

suitable for small scale (100mg–10g) distillation. May be used for vacuum distillattion.

*Practical Hints*
Heating:
Waterbath to 100°C, liquid paraffin to 135°C, silicone oils to 180°C (some will remain stable for a short time up to 250°C), metal baths for higher temperatures (e.g. Wood's metal). Heating mantles are only suitable for distillation of solvents or for heating thermally stable compounds under reflux (they are not easy to regulate and transfer heat irregularly, and are thus dangerous).

Avoid bumping:
Atmospheric pressure
- anti-bumping chips, boiling sticks, platinum tetrahedra, magnetic stirrers.

Vacuum
- capillaries, magnetic stirrers.

Flask size:
About 1.5 times the sample volume.

Columns:
Match the size to the quantity of material; large columns lead to excessive wastage. The efficiency of columns depends on various factors:
– column type

| Column | Diameter mm | Load ml $h^{-1}$ | Theoretical plate height cm |
|---|---|---|---|
| Empty column | 6 | 115 | 15 |
| Vigreux | 12 | 194 | 7.7 |
| Packed column | 24 | 100–800 | 6 |
| Spinning band | 5 | 50–100 | 2.5 |

– column loading (it must not 'choke')
– constancy of pressure
– insulation (cotton wool, asbestos string [dangerous!], silvered vacuum jacket)
– reflux ratio (very important, but only adjustable if special apparatus is used)

Thermometer:
The top of the mercury reservoir should be exactly opposite the side arm leading to the condenser.
Cooling:
Cold water through the condenser. If the distillate solidifies in the condenser use warm water (water bath and circulating pump) or change to an air condenser.
Receiver adaptor:
If it is necessary to change receivers during distillation without breaking the vacuum, a special adaptor should be used (e.g. a 'spider' or 'piglet', or a Perkin triangle).
Receiver:
*Weighed* round-bottomed flasks (essential for measuring yields).
Cold traps:
*Essential* for high-vacuum work (to protect the pump from contamination by volatile substances). N.B. Even at −78°C some solvents may be volatile under high vacuum; pre-drying at room temperature using a water pump usually removes these materials.

Cold bath:
A Dewar flask of solid $CO_2$ in propan-2-ol is effective.
Manometer:
Measures pressure. Not needed for distillation at atmospheric pressure (in which case a drying tube or $N_2$ reservoir may be attached to the

apparatus to protect the products).

Distillation record:

A record should be kept for every distillation showing the weights of all fractions.

Example:

Material to be distilled 20.2g

| Fraction | Bath temp. °C | Vapour temp. °C | Pressure torr | Weight g | $n_D$ | Remarks |
|---|---|---|---|---|---|---|
| 1 | 80 | 55 | 760 | 8.3 | 1.444 | |
| 2 | 60-75 | 22-55 | 18 | 1.1 | 1.416 | mixture |
| 3 | 75-80 | 59 | 18 | 6.6 | 1.409 | |

Azeotropes:

There are certain mixtures of liquids which are not separable by distillation even though the boiling points of the components are sufficiently far apart. Such mixtures give a vapour of *constant* composition at the so-called azeotropic boiling point (usually below the b.p. of any of the components – this gives a 'minimum azeotrope'). This azeotropic mixture (or constant boiling mixture) will continue to be distilled as long as the quantities of the components in the flask remain sufficient. On condensation the azeotrope may separate into two liquid phases.

| Components | | Azeotrope | | | |
|---|---|---|---|---|---|
| | bp (°C) | bp (°C) | Composition (%) Azeotrope | Upper phase | Lower phase |
| Water-<br>chloroform | 100<br>61.2 | 56.3 | 3.0<br>97.0 | 99.2<br>0.8 | 0.2<br>99.8 |
| Water-<br>toluene | 100<br>110.6 | 85 | 20.2<br>79.8 | 0.05<br>99.95 | 99.94<br>0.06 |
| Water-<br>acetonitrile | 100<br>82 | 76.5 | 16.3<br>83.7 | – | – |

The fact that toluene (also acetonitrile, dichloromethane, etc.) on distillation will carrys with it a certain percentage of water can be used to remove water (e.g. from a reaction mixture [azeotropic removal of water]).

**BIBLIOGRAPHY**

(a) *Distillation*

*Organicum,* transl. B. J. Hazzard, Pergamon Press, Oxford, 1973.

*Technique of Organic Chemistry,* A. Weissberger ed., Vol. 4, 2nd edition, 1965.

(b) *Azeotropes*

*Handbook of Chemistry and Physics,* Chemical Rubber Co., Sect. D1.

'Purification and Drying of Solvents', pp. 120–140 in this book.

CHAPTER 4

# Chromatographic Methods

## 1. GENERAL

*Object:* Separation of mixtures

*Basic principles:* Chromatographic separations depend on the differences in the partition coefficients of the components between two immiscible phases. One, the mobile phase, moves relative to the other, stationary phase and transports the substances being separated.

The partition coefficient $K$ of a substance in such a two phase system is given by

$$K = c_s/c_m$$

where $c_s$ = concentration of the substance in the stationary phase
$c_m$ = concentration of the substance in the mobile phase.

The greater the partition coefficient of a substance, the greater its concentration in the stationary phase and the more slowly it moves along the chromatographic system.

*Classification:*

| Mobile phase | Stationary phase | Chromatographic technique |
|---|---|---|
| vapour | solid | gas chromatography (gas-solid chromatography) |
| | liquid | gas chromatography (gas-liquid chromatography, GLC) |
| liquid | solid | adsorption chromatography (liquid-solid partition, liquid-solid chromatography, LSC) |
| | liquid | liquid-liquid partition (partition chromatography, liquid-liquid chromatography, LLC) |

According to the mechanism of retention one can further differentiate chromatography into the following types:

| | |
|---|---|
| Adsorption (normal, reversed phase) | chromatography |
| Liquid-liquid partition | |
| Ion exchange | |
| Gel permeation | |
| Affinity | |

The following sections are devoted to partition between mobile liquids and stationary solid phases.

### 1.1 Adsorption Chromatography

Partition between mobile liquids and solid stationary phases.
The success of such separations depends mainly on the choice of the correct phases. Phases may be divided into:
(a) *Normal phases*
stationary phase: *polar* (kieselgel, alumina, cellulose, etc.)
mobile phase: *nonpolar* → polar (hexane → ether → methanol)
(b) *Reversed phases*
stationary phase: *nonpolar* (modified kieselgels, nylon, polystyrene, etc.)
mobile phase: *polar* (methanol/water/acetonitrile)

*Stationary phases*
Kieselgel
By far the most common stationary phase used by preparative organic chemists to separate mixtures. Kieselgel is dehydrated, highly porous silicic acid, ground to give a particle size of 0.04–0.2 mm, with pore diameters of 50–100 Å and a surface area of 200–400 $m^2$ per gram. The ⩾SiOH groups on the surface give kieselgel a weakly acid character. Care is needed with acid-labile substances.

Alumina
Somwhat basic (pH. 9.5). Neutral alumina is prepared by neutralisation to pH 7.0 followed by activation. Alumina can be deactivated by the addition of water.

| Activity | I | II | III | IV | V |
|---|---|---|---|---|---|
| Water content (% by weight) | 0 | 3 | 6 | 10 | 15 |

Preparation: Add the calculated amount of water. Shake until all lumps and damp patches disappear. Allow to stand *ca.* 24 h in a tightly stoppered vessel.

*Classification*

| *Substance* | *Eluent* | *TLV (ppm)* |
|---|---|---|
| saturated hydrocarbons | *n*-pentane, *n*-hexane | 100 |
| unsaturated hydrocarbons | cyclohexane | 300 |
| | carbon tetrachloride | 10! |
| | toluene | 200 |
| ethers | benzene (carcinogen) | 10! |
| | diethyl ether | 400 |
| esters | chloroform (possible carcinogen) | 10! |
| ketones | dichloromethane | 200 |
| amines | acetone (not on $Al_2O_3$) | 1000 |
| | ethyl acetate | 400 |
| alcohols | *iso*-propanol | 400 |
| | ethanol | 1000 |
| phenols | methanol | 200 |
| acids | acetic acid | 10 |
| | water ↓ | – |
| increasingly stongly adsorbed on kieselgel and alumina | increasing eluting strength ('eluotropic' series) | |

Most mixtures of solvents can be used as eluents.

Only use solvents of known purity; they should normally be freshly distilled (ethers form peroxides, chloroform may contain alcohol or phosgene, solvents pick up moisture, etc).

*Temperature dependence*

The lower the temperature, the more stongly substances are adsorbed on the stationary phase: carry out chromatography in an area which is draught-free and not too hot. In favourable cases, cooled columns may give improved separation, and may be advantageous for the separation of thermally labile compounds.

## 2. ANALYTICAL AND PREPARATIVE THIN-LAYER CHROMATOGRAPHY

### 2.1 Thin-layer chromatography (t.l.c.)

(a) *Uses*

- Checking purity.
- Preliminary tests before separation.
- Qualitative comparison with known substances.
- Keeping check on reactions.

(b) *Procedure*

(i) Cut commercial glass plates (20 × 20 cm) with a glass cutter (wheel, rather than diamond).

(ii) Choose stationary phase – alumina or kieselgel of *ca.* 0.2 mm thickness.

(iii) Apply the substance (or mixture) – about 2μl of a dilute (1%) solution of the substance in the least polar suitable low boiling solvent is applied to the plate in the form of a spot and the solvent allowed to evaporate completely.

(iv) Develop the chromatogram – The plate is dipped into the eluent and allowed to develop. When the solvent front has advanced a suitable distance, the plate is removed (see Fig.), the solvent front marked, and the plate allowed to dry.

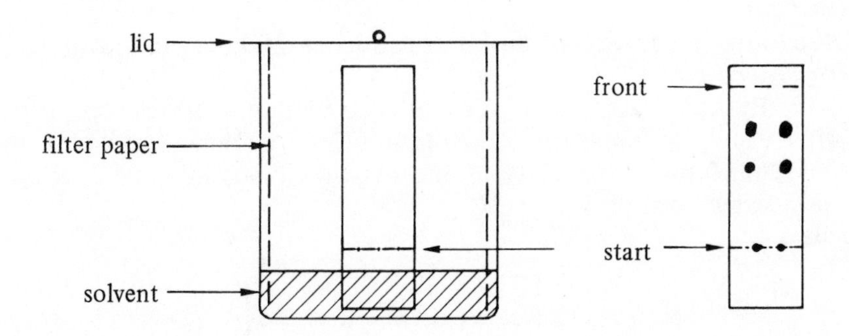


(v) Detection – coloured spots are, of course, immediately visible. Colourless spots can be made visible by:

(1) UV: if the substance absorbs UV, one can use a stationary phase impregnated with a fluorescent indicator.
(2) standing the plate in iodine vapour.
(3) spraying with conc. $H_2SO_4$-water 1:1 (in a specially protected compartment in a fume hood!!) and then heating strongly e.g. with a hot-air blower to carbonise the compounds.
(4) spraying with suitable colour reagents. For a comprehensive list see Ref. [1].

(c) *Recording*

Trace the chromatogram using tracing paper. Mark in spots, starting positions, and solvent front and record type of plate, eluent, and method of development.

$$R_f = \frac{\text{Distance of spot centre from start}}{\text{Distance of solvent front from start}}$$

The $R_f$ value depends on the conditions under which the chromatogram was run (type of plate, eluent, etc., as well as condition of plate, temperature, saturation of vapour, etc.). Its reproducibility is about ± 20%. It is best to run probable reference compounds on the same plate.

(d) *Multiple runs*

It is sometimes useful to re-run a plate, after it has dried, until the solvent front reaches the position of the solvent front on the previous run. An $n$-fold re-running is effectively the same as running a chromatogram $n$-times the length, and gives better resolution for very close spots of low $R_f$.

## 2.2 Preparative TLC (thick-layer chromatography)

(a) *Uses*

Preparative separation of mixtures (up to *ca.* 200 mg practicable).

(b) *Procedure*

(i) By t.l.c. optimise the system (adsorbent, eluent) for separation.
(ii) Apply a concentrated solution (*ca.* 100 mg $cm^{-3}$) of the mixture to the plate as the narrowest possible strip, using a drawn-out pipette.

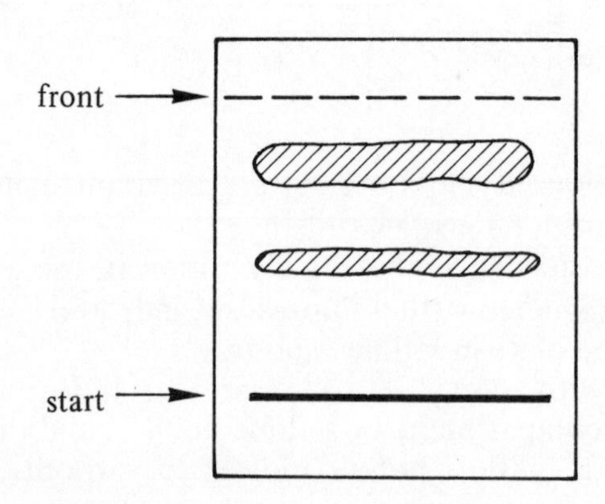


(iii) Elute as for t.l.c.
(iv) Using a UV lamp if necessary, locate the individual bands and mark them with a pencil. (Alternative methods: cover the plate carefully with aluminium foil, leaving a 0.5 cm strip in the middle exposed and develop in iodine vapour – then assume the bands are straight!! Especially if plastic or aluminium

backed plates are being used, a narrow marker strip can be cut from the centre and the bands located by one of the methods discussed in the section on analytical t.l.c.).

(v) Scrape off the individual bands.

(vi) Elute the substance with a solvent more polar than the eluent. *Warning:* Kieselgel is slightly soluble in ethanol or methanol. If one of these solvents has to be used, evaporate the alcohol solution after filtration, take up the residue in a less polar solvent, and filter again.

(vii) Evaporate the solvent. Crystallise or distil the product and characterise it.

*Note:* Many plates are slightly contaminated by plasticisers or binding agents. For small-scale work plates should be cleaned by first running in a mixture of chloroform and methanol and then allowed to dry thoroughly before the mixture is applied. It is also essential to use pure, redistilled solvents for all processes.

## 3. COLUMN CHROMATOGRAPHY

(a) *Uses*

Separation of larger quantities of material (e.g. > 100 mg).

(b) *Procedure*

(i) By t.l.c. optimise the system (adsorbent, eluent) for separation (see also section 3(c)).

(ii) Quantity of adsorbent required: 30–100 times the weight of the mixture (50 times is a fair average).

(iii) Column dimensions: the adsorbent should pack into it with a 10:1 ratio of height:diameter. To estimate the quantity of adsorbent in relation to the column diameter, See pp. 42–44.

(iv) Securely clamp column vertical.

(v) Fill the column with the eluent (or, if a mixed solvent is being used, with the less polar component). For choice of solvents, see section 3(c).

(vi) Press a plug of cottonwool firmly into the bottom with a glass rod. Cover it with a layer of clean sea-sand (*ca.* 0.5–1 cm).

(vii) Pour in the weighed absorbent:
Alumina – in a fine stream while tapping the column with a piece of rubber tubing.
Silica gel – as a slurry in the solvent.

(viii) Allow the solvent to run out (tapping the column) until *ca.* 1 mm remains above the top of the adsorbent.

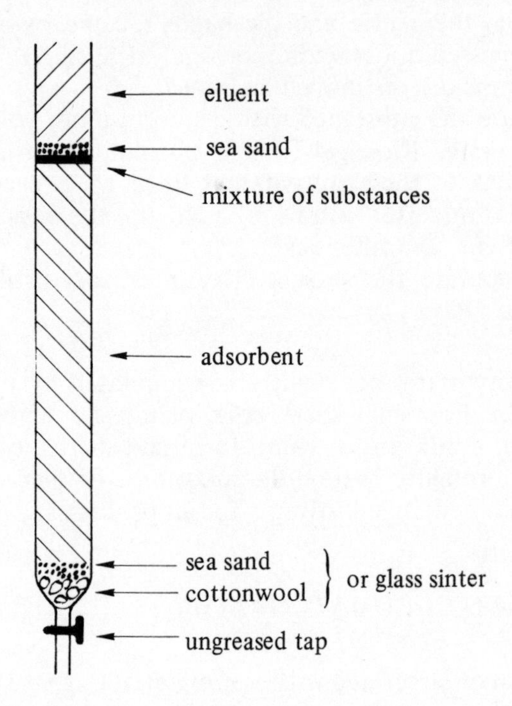


(ix) Dissolve the substance in the minimum amount of solvent (not more polar than the eluent) and apply carefully and evenly to the top of the column (tap closed). Open the tap and run off solvent until a 1 mm layer again remains above the adsorbent. Repeat the process a few times using small amounts of eluent.

(x) Fill the volume above the column of adsorbent carefully with eluent. To protect the surface of the adsorbent, a 1 cm layer of sea-sand can be deposited on the top of the column. For choice of solvent, see section 3(c).

(xi) Open the tap and collect the eluent in fractions (*ca.* $x$ ml fractions for $x$ g of absorbent).

(xii) Flow rate: poor separation results at high flow rates. Rule of thumb: 3–4 ml per minute with a column 40 cm high.

(xiii) Evaporate each fraction and weigh the residue. Enter the information on the record (see below).

(xiv) Fractions containing the same components (check by t.l.c.) can be combined, purified (crystallisation, distillation) and characterised.

thin layer chromatogram of individual fractions

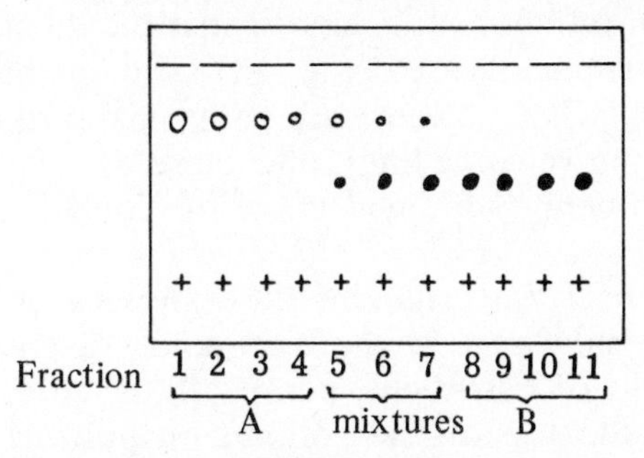


(c) *Most common eluents (for kieselgel and alumina columns)*
The choice of eluent depends on the polarity of the substances to be separated. One often needs to use mixtures of solvents. In general, one begins with petroleum ether, dichloromethane, ethyl acetate, and mixtures of these. Dichloromethane and ethyl acetate are also good solvents for many substances.

| *Common eluents* | *Volume Ratios often employed* |
|---|---|
| | ∞ ↓ |
| petroleum ether | |
| | 9:1 ↓ |
| toluene (rather than benzene) | |
| | 4:1 ↓ |
| diethyl ether<br>*t*–butyl ether | |
| | 1:1 ↓ |
| dichloromethane | |
| | ∞ ↓ |
| ethyl acetate | |
| | 9:1 |
| methanol | |

The minimum quantity of a solvent mixture used should be such that the solvent front reaches the end of the column before a mixture of different composition is applied.

USEFUL TIPS:
– Do not let the column run dry!
– The temperature around the column should be kept approximately constant.
– A chromatographic run should be completed without interruption.

– When using alumina, do not have acetone as an eluent – you may find condensation products as the major constituents of the eluate!

– When low-boiling eluents are being used (ether, pentane, dichloromethane), especially when solvents are being changed, bubbles often appear in the column (channelling), and these usually prevent satisfactory separation being obtained. Use of cooled columns may prevent channelling.

For difficult separation problems the following procedure often works: dissolve the substance in the least polar solvent possible, add 1–2 times its weight of adsorbent, pump off the solvent, and settle dry powder onto the top of the column by pouring it through a small layer of eluent. Then elute as usual.

For column chromatography it is often preferable to use an eluent mixture slightly less polar than the one shown to be the best for t.l.c.

C) *Diagrams for estimation of column dimensions and quantities of adsorbent for column chromatography*

The following diagrams enable one to estimate the amounts of silica gel (Diagram 1) or alumina (Diagram 2) required to fill a chromatgraphic column to a particular height.

*Example A.* How much silica gel is required to fill about 4/5 of a column of diameter 18 mm and length 57 cm?

One Diagram 1: for an 18 mm diameter, 26 g of silica gel (packed as a suspension) will give 18 cm (diam.:height = 10:1!) of packing.

Therefore, for 45.6 cm (4/5 of 57) one requires 66 g (45.6/18 × 26) of silica gel.

*Example B.* A mixture requires 110 g of alumina for chromatographic separation. How large a column is required?

From Diagram 2: 110 g alumina requires a column, diameter 24 mm and height > 24 cm. If one wishes to use a column whose diameter is 17 mm, calculate the height as in Example A. [110 g alumina will fill 49 cm].

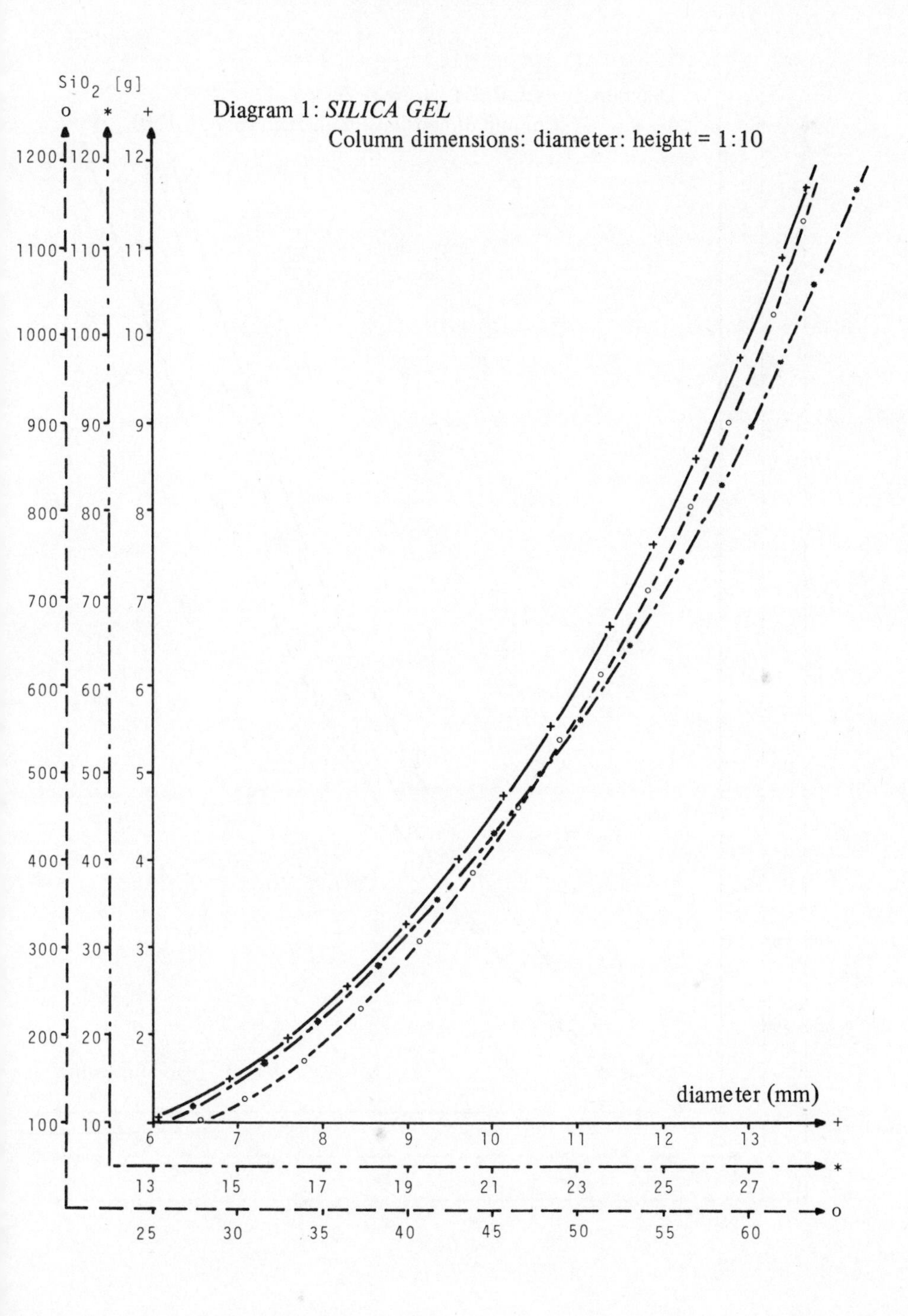
$SiO_2$ [g]
o * +
Diagram 1: *SILICA GEL*
Column dimensions: diameter: height = 1:10
1200 120 12
1100 110 11
1000 100 10
900 90 9
800 80 8
700 70 7
600 60 6
500 50 5
400 40 4
300 30 3
200 20 2
100 10
diameter (mm)
6 7 8 9 10 11 12 13 +
13 15 17 19 21 23 25 27 *
25 30 35 40 45 50 55 60 o

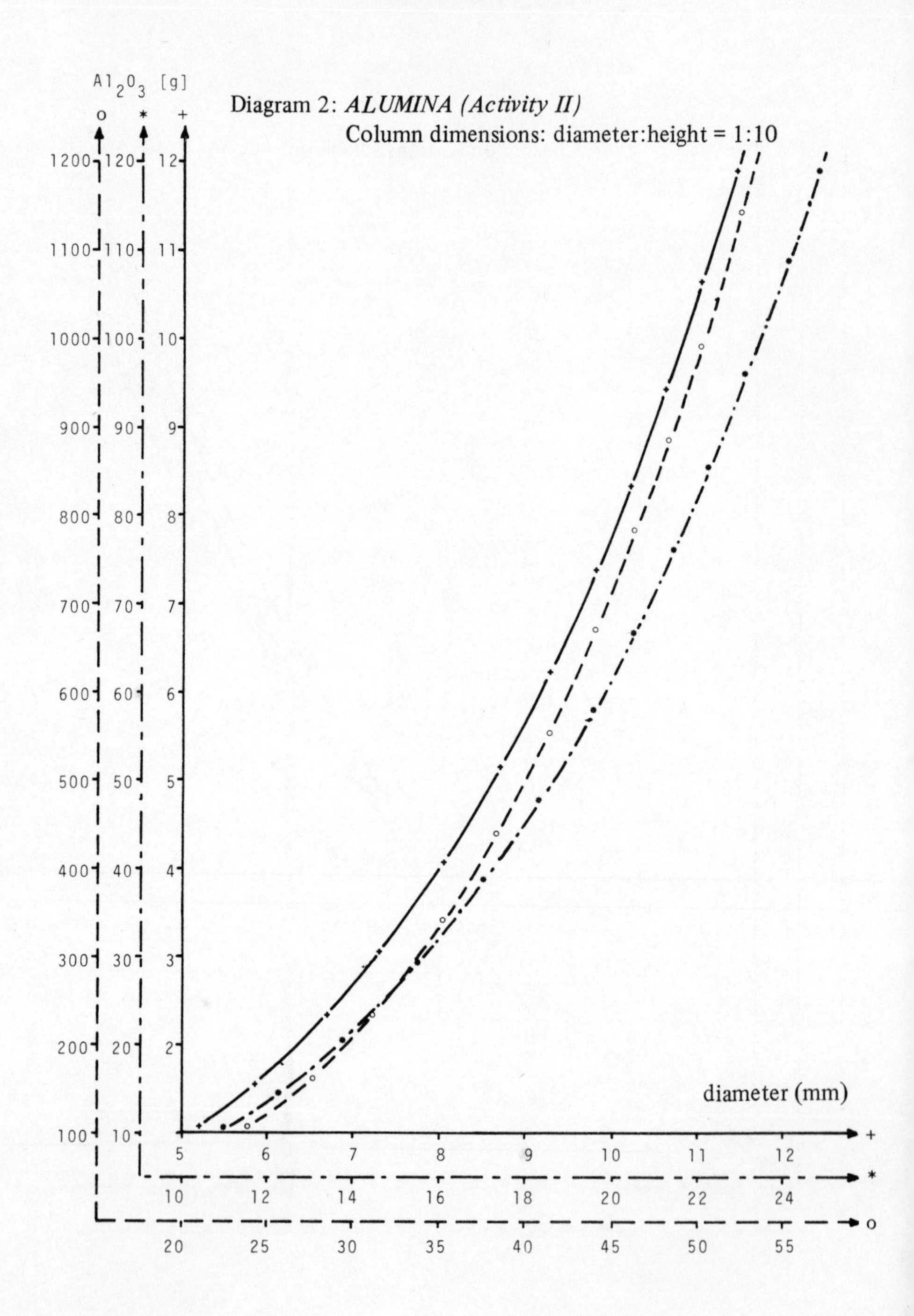

$Al_2O_3$ [g]
o
*
+
Diagram 2: *ALUMINA (Activity II)*
Column dimensions: diameter:height = 1:10
1200 120 12
1100 110 11
1000 100 10
900 90 9
800 80 8
700 70 7
600 60 6
500 50 5
400 40 4
300 30 3
200 20 2
100 10
diameter (mm)
5 6 7 8 9 10 11 12
10 12 14 16 18 20 22 24
20 25 30 35 40 45 50 55

CHROMATOGRAM

No. AM-7 Start: 09.30 Finish: 11.30

Substance: Reaction mixture Weight: . . . g

Alumina (Woelm) a ⓝ b / Activity (III)
Kieselgel (. . .) a n / Mesh ( ) } 10 g 55 fold amount

Adsorbed from: dichloromethane 1 $cm^3$

Column dimensions: diameter: 1.5 cm height: 17 cm

| Fraction | Solvent | Ratio | Vol ($cm^3$) | Flask (g) | Wt (mg) | Total wt (mg) | Analytical data etc. |
|---|---|---|---|---|---|---|---|
| 1 | hexane/ | 1:1 | 10 | 66.345 | 0 | 0 | |
| 2 | $CH_2Cl_2$ | 1:1 | 10 | 47.407 | 0 | 0 | |
| 3 | " | 1:1 | 10 | 52.578 | 0 | 0 | |
| 4 | " | 1:1 | 10 | 38.598 | 1.8 | 1.8 | Combined: |
| 5 | " | 1:1 | 10 | 36.189 | 35.1 | 36.9 | tlc 17 |
| 6 | " | 1:1 | 10 | 39.857 | 17.5 | 54.4 | IR 12 |
| 7 | " | 1:1 | 10 | 47.425 | 5.2 | 59.6 | UV 5 |
| 8 | " | 1:1 | 10 | 44.577 | 1.2 | 60.8 | |
| 9 | " | 1:1 | 10 | 46.071 | 0 | 60.8 | |
| 10 | " | 1:5 | 20 | 35.346 | 0 | 60.8 | |
| 11 | " | 1:5 | 20 | 50.634 | 0 | 60.8 | |
| 12 | " | 1:5 | 30 | 45.514 | 0 | 60.8 | |
| 13 | $CH_2Cl_2$ | | 10 | 62.080 | 5.0 | 65.8 | Combined: |
| 14 | " | | 10 | 52.578 | 11.2 | 77.0 | Alc 18 |
| 15 | " | | 10 | 59.167 | 31.3 | 108.1 | IR 13 |
| 16 | " | | 10 | 54.628 | 3.5 | 111.6 | UV 6 |
| 17 | " | | 10 | 42.407 | 2.1 | 113.7 | |
| 18 | $CH_2Cl_2$/ | 1:1 | 20 | 53.212 | 3.2 | 116.9 | |
| 19 | EtOAc | 1:3 | 20 | 42.343 | 5.7 | 122.6 | |
| 20 | " | 1:3 | 50 | 36.189 | 4.0 | 126.6 | |
| 21 | " | 1:3 | 50 | 33.432 | 1.6 | 128.2 | |
| 22 | | | | | | | |

## 4. RAPID COLUMN CHROMATOGRAPHY

('Flash chromatography') (Ref [2])

(a) *Application*

Separation of 100 mg – 10 g of mixtures of substances

- containing few components (2–3)
- components differ in $R_f$ on t.l.c. > 0.15

(b) *Procedure*

– By thin layer chromatography (kieselgel) select an eluent which gives the $R_f$ of the desired component *ca* 0.35. If several components are to be isolated the average $R_f$ should be 0.35; however, if the components are widely separated 0.35 should be the $R_f$ of the component which moves most slowly.

– Use a column with height: diameter ratio = 10:1, fitted with a pressure head and with a small dead-volume at the bottom.

– Pack the column dry with kieselgel (40–63 μm mesh) 100–200 times the weight of the mixture to be separated. The column should be no more than 50% full.

– To protect the surface of the adsorbent and to improve separation between different solvents, add a 1 cm layer of fine sand.

– Fill the column with solvent and force it through the kieselgel using *ca.* 0.4 atm. pressure of the gas (air, $N_2$, Ar). Adjust the valve to give a flow rate of *ca.* 5 $cm^3$/min. *Care!*

- Dissolve the mixture in a little solvent, apply to the column, and allow it to percolate through the sand layer.
- Fill the column with eluent.
- Replace the head and apply the gas pressure.
- Collect fractions, 30 sec-1 min each.
- Evaporate fractions, weigh, and test for homogeneity by t.l.c.
- Combine identical homogeneous fractions and characterise.

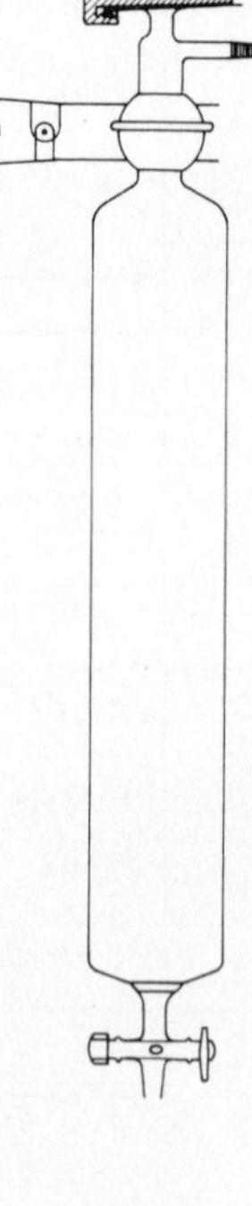

(c) *Practical Tips*

– For safety, cover the column with protective (polypropylene) mesh.

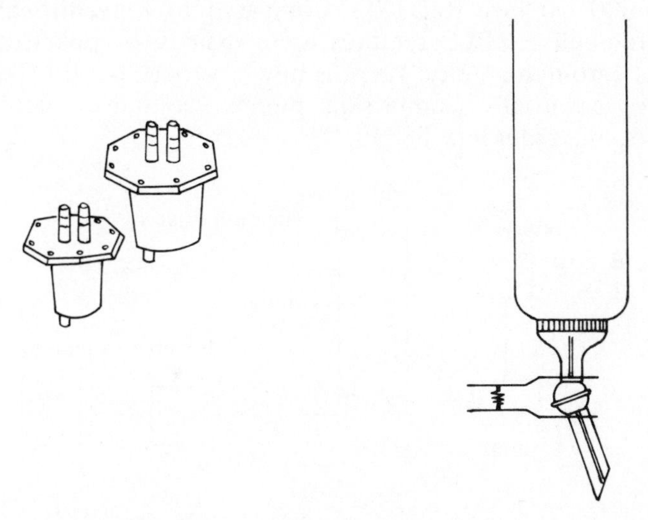

– Any column with B29 (or B34 or B45) joints and a small dead-space can be used for flash chromatography, if fitted with a commercial polypropylene stopper with two connecting nozzles as pressure head: the flow rate is regulated by means of a piece of tubing fitted with a clamp. It is useful to have a ball-and-socket ground glass joint (diam. 12.7 mm, bore 2 mm) at the foot of the column: the column is shut off by moving the socket to the side.

– If ether is used as eluent, the heat of interaction with the silicagel causes bubbles to appear in the adsorbent, and these are difficult to eliminate. In this case it is better to pack the column with a slurry of silicagel.

– Repeated use of the same column: after use, the column can be washed and then re-used. First allow two column volumes of methanol to flow through, then similar volumes of ethyl acetate and hexane (or the solvent mixture best suited to the next chromatogram). The retention time (or 'activity' of the column) is only slightly reduced by this procedure.

**HPLC**

The acronym is said to stand for high pressure liquid chromatography or for high-performance liquid chromatography.

This efficient separatory technique has become widely used since 1970. It can be used for analytical and preparative separations on a small scale ($\mu$g to *ca.* 100 mg). The high efficiency is achieved by the use of stationary phases with very fine, uniform mesh (typically 3–10 $\mu$m diameter) packed with a slurry of stationary phase at a pressure of 300–600 bar (see Ref [3]). Compared to conventional column chromatography, HPLC requires more expensive apparatus and, as with gas chromatography, there is now a variety of HPLC apparatus available, essentially comprising pumps, columns, inlet systems, detectors, and recorders. See Fig.

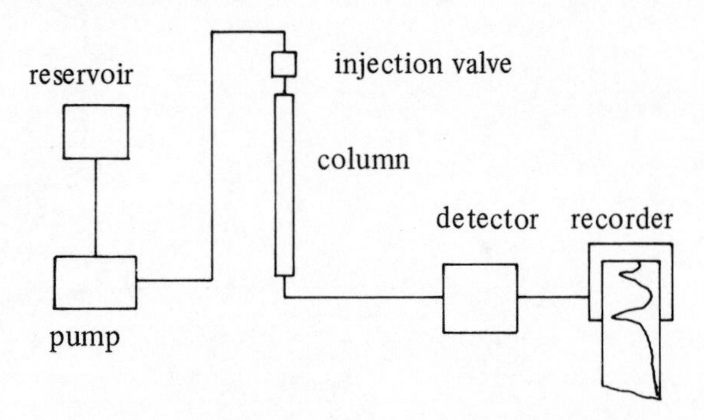


In addition one can obtain a variety of stationary phases: various kieselgels (e.g. 5 ± 2 $\mu$m mesh), modified kieselgels, aluminium oxide, and materials for reversed-phase HPLC. The most common detectors employ refractometric or spectrophotometric techniques.

The choice of particular separation systems depends on the nature of the substances to be separated. The following table may serve as a useful guide.

| **Class of Substance** | **Separation Method** |
|---|---|
| Lipophiles, neutral organic compounds with MW <2000 | normal adsorption |
| hydrophiles, polar organic compounds, saturated hydrocarbons, MW <5000 | reversed phase adsorption |
| moderately polar organic compounds | (liquid-liquid) partition |
| acids, bases, zwitterions, MW <10 000 | ion exchange |
| separation by molecular size of all sorts of compounds | gel permeation |

In the following section, we assume that an efficient HPLC with a packed kieselgel column and fitted with a detector is available. A mixture of neutral organic substances is to be separated.

(a) *Procedure:*
– Determine the UV/visible spectrum of the mixture.
– Consider only those solvents transparent in the wavelength regions where the compounds absorb. The most useful solvents for HPLC are: hexane (200†), diethyl ether (220), *t*-butylmethyl ether (220), dichloromethane (230), ethyl acetate (253), chloroform (242), acetonitrile (200), methanol (210), water.
– Use t.l.c. to choose the best solvent (or mixture of solvents). The $R_f$ values for the substances to be separated should be about 0.3. If the $R_f$ reaches 0.9–1 even in hexane, kieselgel is not a suitable stationary phase; reversed phase HPLC may be considered.
– To obtain consistently high separation efficiency the chosen solvent should be prepared as follows:

(a) Use solvents only of known purity, HPLC quality if necessary.

(b) To obtain a defined water content in the mobile phase, mix solvent saturated with water and absolute solvent in a known proportion (e.g. 1:1).

(c) Remove any dust particles from the solvent by filtration through a sinter ($< 1$ μm pore size).

(d) De-gas the solvent: apply moderately reduced pressure or use ultrasonics.

– Condition the column with the chosen solvent (or mixture) until the detector and recorder show a constant base-line. Large, sudden changes in solvent polarity are to be avoided.
– Apply a small quantity (∿0.1 mg) of the mixture, dissolved in the eluting solvent, to obtain a test chromatogram. Adjust the polarity of the mobile phase to obtain a reasonable retention ($k' \sim 3$). If no separation results, alter the polarity to increase $k'$ ($k' \to 10$). If there is still no separation, change to a mobile phase composed of other solvents.
– For reliable separations, one must ensure that the capacity factor $k'$ and the resolution $R_s$ (see p. 50) remain constant each time the substance is applied.
– Collect fractions giving the same signal.
– After concentration of each component, test for purity by applying once more.

For separation on reversed phase columns, water-methanol, water-acetonitrile, or water-tetrahydrofuran mixtures are used. If the apparatus has no facilities for gradient elution, test first using a 1:1

† The figure is the wavelength in nm at which the transmittance of a 1 cm layer of the solvent is $< 10\%$.

solvent mixture. If $k'$ is too low, increase the proportion of water; if the substance is not eluted at all (i.e. $k'$ too large), increase the proportion of the organic solvent.

(b) *Some additional concepts, plus the calculation of the number of theoretical plates of a column* (including connections and detector cell):

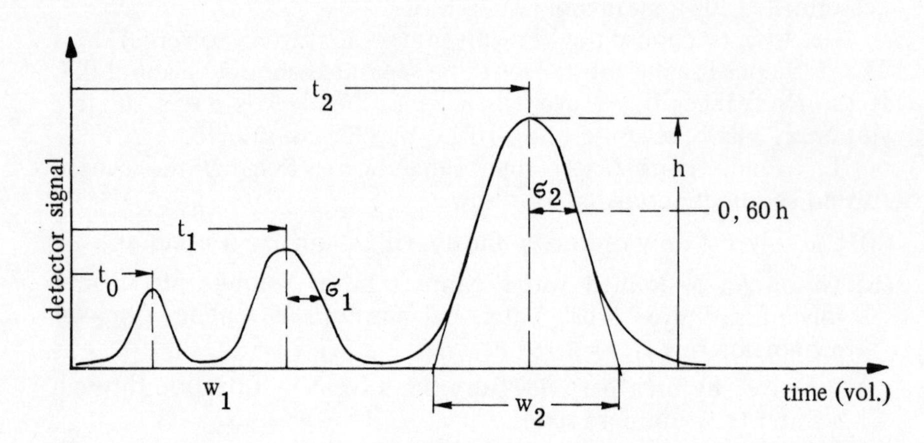


| | | |
|---|---|---|
| $t_0$ | = | time between injection and appearance of undelayed material ('solvent front') at detector |
| $t_1, t_2$ | = | time for appearance of 1st and 2nd substances at detector |
| $h$ | = | signal height |
| $\sigma$ | = | half-width of signal at 0.606 h = standard deviation for Gaussian distribution |
| $W$ | = | signal width, determined from tangents to the sides of the signals |
| $N$ | = | no. of theoretical plates |
| | = | $(t_1/\sigma_1)^2$ (for 1st signal). |
| $H$ | = | plate height (more accurately: equivalent height per theoretical plate) |
| | = | $L/N$ (where $L$ = column length) |
| $k'$ | = | capacity factor (measure of the retention of a substance, *cf.* $R_f$ in t.l.c.) |
| | = | $(t_1 - t_0)/t_0$<br>$k'$ is $\geqslant 0$ |
| $R_s$ | = | resolution of two signals<br>$(t_2 - t_1)/½(W_1 + W_2)$ |

## REFERENCES

[1] *Dyeing Reagents for Thin Layer and Paper Chromatography,* E. Merck, Darmstadt, 1971.

[2] W. C. Still., M. Kahn, and A. Mitra, *J. Org. Chem., (1978)* **43** 2923.

[3] V. Meyer, *Praxis der Hochleistungs-Flussigchromatographie,* Diesterweg-Salle-Sauerlander, Frankfurt am Main, 1979.

## BIBLIOGRAPHY

E. Heftman, *Chromatography,* Reinhold, New York, 1961.

L. R. Snyder & J. J. Kirkland, *Introduction to Modern Liquid Chromatography,* Wiley-Interscience, New York, 2nd ed., 1979.

J. M. Miller, *Separation Methods in Chemical Analysis,* Wiley-Interscience, New York, 1975.

P. A. Bristow, *Liquid Chromatography in Practice,* hetp, Wilmslow, 1977.

*Handbook of Chromatography,* Chemical Rubber Co., Cleveland, 1972.

J. A. Dean, *Chemical Separation Methods,* Van Nostrand Reinhold, 1969.

*Chromatography,* D. R. Browning ed., McGraw Hill, 1969.

J. G. Kirchner, *Technique of Organic Chemistry,* Vol. 12, Interscience, 1967.

H. Engelhardt, *High Performance Liquid Chromatography* (transl, G. Gutnikov), Springer Verlag, New York, 1974.

E. L. Johnson & R. Stevenson, *Basic Liquid Chromatography,* Varian, Palo Alto, 1978.

CHAPTER 5

# Extraction and Isolation

Mixtures, especially those from reactions, can often be separated at little expense by extraction.

Extraction can be defined as the transfer of a substance X from one liquid phase, A, into an immiscible liquid phase, B. The partition of X between the immiscible liquids A and B is given by the Nernst Partition Equation.

$$C_B(X)/C_A(X) = K_T$$

where $C_A(X)$ and $C_B(X)$ are the concentrations of X in A and B respectively.

$K_T$ = the partition coefficient at temperature $T$.

If X is much more soluble in B than in A ($K \geqslant 100$), 2–3 extractions will suffice to 'shake out' X. Conversion of a substance into its salt alters its partition coefficient drastically. By this means, organic molecules may be rendered water-soluble and, as we see below, can then be separated easily from other (neutral) substances by shaking out.

If $K \leqslant 100$, repeated extraction with fresh portions of B will be necessary. In practice, one uses special apparatus for repeated extractions (counter-current distribution – Craig apparatus) or for continuous extraction (using immiscible solvents either more dense or less dense than water).

### Separation of mixtures according to acidity

This variant of extraction is one of the commonest processes in preparative organic chemistry. The following procedure is therefore particularly suited to preparative experiments.

*Principle*
Carboxylic acids and phenols are deprotonated by bases (sodium bicarbonate, and sodium carbonate or hydroxide, respectively) and the anions are then water-soluble. Amines are rendered water-soluble by protonation in acidic media.
*Advantages*
Rapid separation and consequent safer handling of the relevant substances, especially as one often shakes them out using cold aqueous solutions (addition of ice).
*Disadvantages*
The solvents must be removed subsequently. Also, the method is not suitable for compounds that react with water.
*Procedure*
(a) Preliminaries
– Prepare requisite aqueous solutions
(i) Saturated $NaHCO_3$ (95.7 g $\ell^{-1} \sim 1N$)
(ii) Saturated $Na_2CO_3$ (211.97 g $\ell^{-1} \sim 2N$)
(iii) $2N$ NaOH (80 g $\ell^{-1}$).
(iv) $2N$ HCL (200 $cm^3$ conc. $\ell^{-1}$).
(v) Saturated NaCl.
– Have $H_3PO_4$ ready for acidification of basic extracts ($2N$ $H_2SO_4$ as second choice).
– Moisten separatory funnel taps with water (or, if necessary, grease lightly); make sure they do not leak!
– Have a supply of ice available.
– Make sure there are no naked flames in the vicinity.
(b) Separation
– Dissolve mixture in ether. Filter if necessary (and in that case triturate the residue three times with ether, filter off the solvent, and combine the ethereal extracts). The ether insoluble material should be tested to see whether it dissolves in water.
– Put ice into 3 separatory funnels.
– Put some ether into the second and third funnels.
– Put the ethereal solution into the first funnel.
– During the following operations the ether layers remain in their respective funnels. Funnels should be no more than two-thirds full at any time. Hold the stoppers and taps in place while shaking. Shake carefuly at first, and release internal pressure via the tap (invert the funnel!) repeatedly until no excess pressure remains, then shake vigorously about 20 times.
(i) Add a portion of saturated $NaHCO_3$ solution to the first funnel; shake. Transfer aqueous layer to second funnel, shake; transfer to third funnel; shake. Run off aqueous layer.

Separation of organic substances

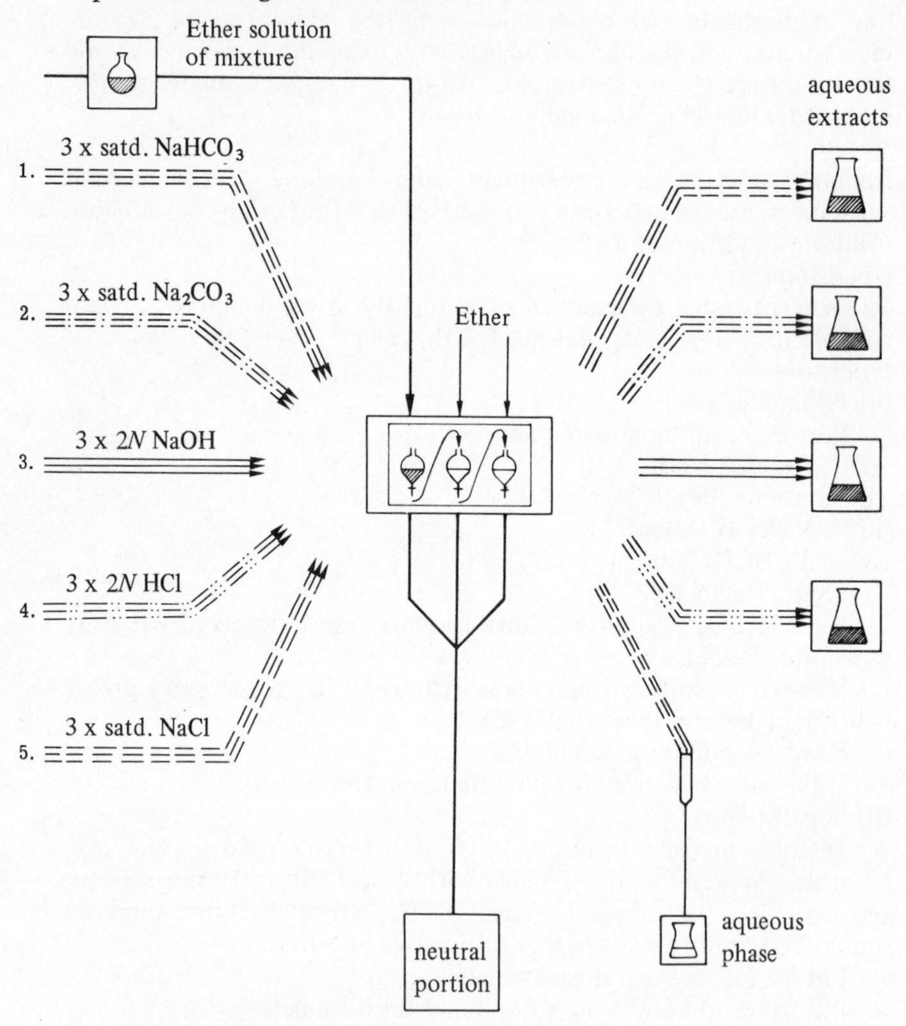


(ii) Repeat the process with two more portions of saturated $NaHCO_3$. Combine the $NaHCO_3$ extracts.

(iii) Now (as in Scheme A) continue the extraction procedure in the same way using the other aqueous solvents (*viz.* saturated $Na_2CO_3$, 2*N* NaOH, 2*N* HCl).

(iv) In the final stage, extract the ether layers with saturated NaCl until the aqueous layer is neutral. (It may be necessary to add a trace of $NaHCO_3$.

Extraction of organic substances from aqueous phase

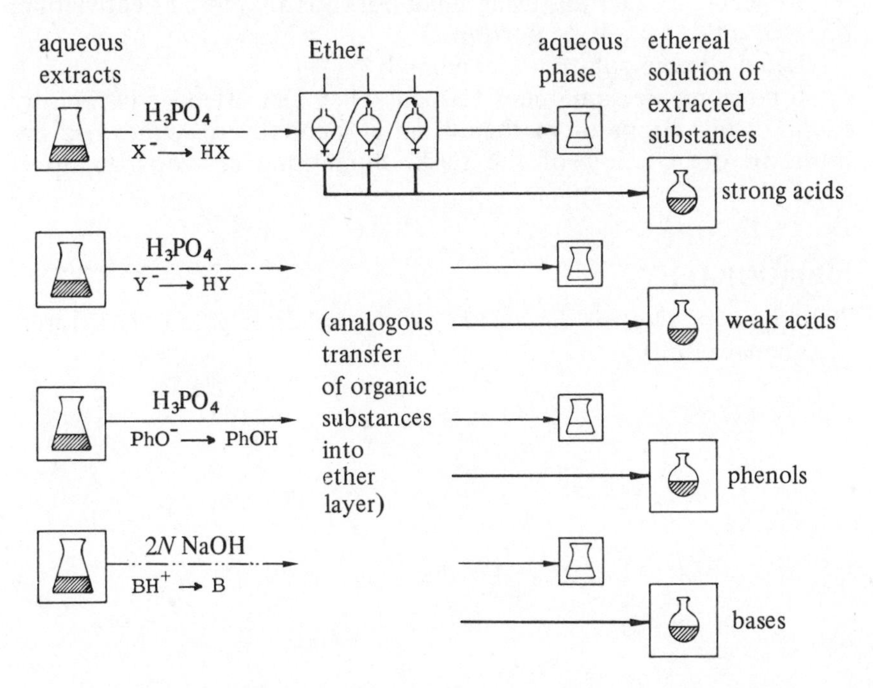


(c) Extraction of organic substances from the aqueous layers

– Work up the basic layer containing phenolic salts first (they are the most likely to decompose).

– Each basic aqueous layer is mixed with ice and then carefully acidified to pH 1–2 in a large beaker using $H_3PO_4$ (*Careful:* $CO_2$ is evolved).

– Ice is added to the HCl extract which is then basified to pH ∿ 10 with 2*N* NaOH.

– For the extraction of each of the four aqueous solutions, they are placed in equal portions in three separatory funnels, and shaken with ice and ether.

– The three ethereal extracts from each solution are combined, dried over anhydrous $MgSO_4$, except that the bases should be dried over anhydrous $K_2CO_3$ (and acid-sensitive substances over $Na_2SO_4$), filtered, the filtrates evaporated, and the residues dried to constant weight.

– The aqueous layers should be saved until one has accounted for the material.

(d) Some hints

– Work quickly (Separatory funnels are not for storing solutions!)

– Repeated extractions using small portions are more effective than one extraction by a large portion.

– Label all your solution immediately!

– Emulsions are sometimes formed. They can often be broken by adding a few drops of methanol, amyl alcohol, etc. Otherwise try filtration or rotation of the flask. Sometimes all you can do is wait!

## BIBLIOGRAPHY

*Technique of Organic Chemistry,* A. Weissberger ed., Vol. 3, Interscience, 1950.

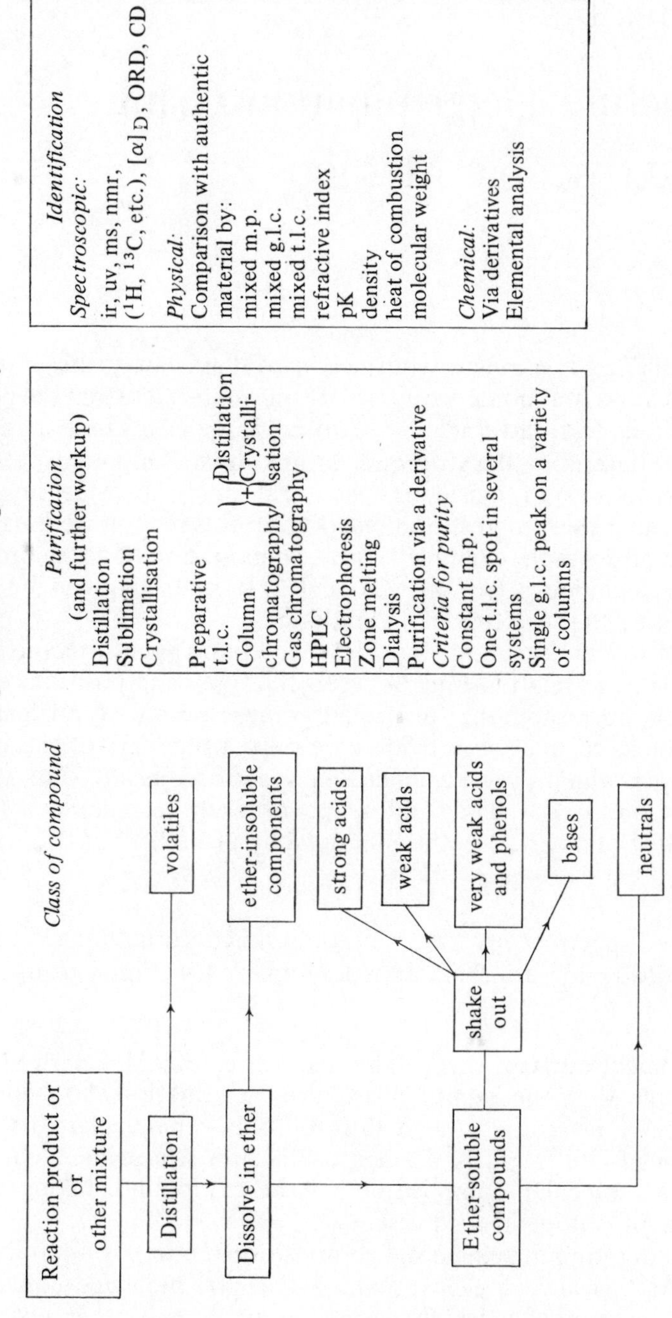

Outline of Work up Procedures in Organic Chemistry
Class of compound
Reaction product or
or
other mixture
Distillation
volatiles
Dissolve in ether
ether-insoluble
components
Ether-soluble
compounds
shake
out
strong acids
weak acids
very weak acids
and phenols
bases
neutrals
Purification
(and further workup)
Distillation
Sublimation
Crystallisation
Preparative
t.l.c.
Column
chromatography
+
Distillation
Crystalli-
sation
Gas chromatography
HPLC
Electrophoresis
Zone melting
Dialysis
Purification via a derivative
Criteria for purity
Constant m.p.
One t.l.c. spot in several
systems
Single g.l.c. peak on a variety
of columns
Identification
Spectroscopic:
ir, uv, ms, nmr,
(¹H, ¹³C, etc.), [α]D, ORD, CD
Physical:
Comparison with authentic
material by:
mixed m.p.
mixed g.l.c.
mixed t.l.c.
refractive index
pK
density
heat of combustion
molecular weight
Chemical:
Via derivatives
Elemental analysis

CHAPTER 6

# Structure Determination Using Spectroscopic Methods

Instrumental methods are routinely used these days to assign structural features and functional groups to organic molecules from the positions, intensities, and patterns of spectroscopic signals. These data may be used to determine the structures of unknown compounds, to confirm the structures of synthetic products, or to provide information concerning the progress of a reaction. In this section we illustrate, by means of a simple example, how a combination of common spectroscopic techniques (ir, ms, $^{1}H$ and $^{13}C$ nmr, uv) can be used to determine the structure of a compound.

All recent textbooks include chapters on spectroscopic methods in which the relationships between the types and positions of signals and the corresponding functional groups and structural features are explained, so that we provide here only a summary of the structural elements which may be identified spectroscopically. For a detailed discussion of the basis and scope of each technique, consult the specialist texts listed in the Bibliography at the end of this section.

**Infrared spectroscopy** (ir): Vibrations and deformations in the range 4000–200 $cm^{-1}$ are characteristic of many functional groups.

**Mass spectrometry** (ms): The radical cation $M^{+\cdot}$ derived from a molecule M in the vapour phase gives information about the relative molecular mass and – at high resolution – molecular formula. The molecular ion ($M^{+\cdot}$) is defined as the ion whose mass corresponds to the molecular composition of the compound, taking the most abundant isotope in each case.

In determining elemental composition it is important to remember that $M^{+\cdot}$ is always even provided that the molecule contains only atoms whose valencies and atomic weights are either both even or both odd (H, C, O, S, Si, Cl). For other elements or isotopes, such as N, $^{13}C$, $^{2}H$, the value of $M^{+\cdot}$ is odd unless the number of such atoms is even. In addition, fragmentation of $M^{+\cdot}$ must involve the loss only of chemically reasonable fragments. Accordingly, mass differences of 3–13, 21–24, 37, 38 are not reasonable.

Besides the molecular ion, peaks are observed which derive from natural isotopic abundances. From the (M+1) peak one may estimate the maximum number of C atoms; (M+2) is diagnostic for e.g. Cl, Br, Si, S.

**Nuclear magnetic resonance spectroscopy** (nmr): Certain nuclei, such as $^{1}H$, $^{13}C$, $^{19}F$, possess a magnetic moment which can adopt parallel or antiparallel orientations in a magnetic field. Transitions between the energy levels associated with these two orientations are induced by radio frequency electromagnetic radiation. The measured energy difference depends, among other factors, on the chemical environment of the particular nucleus, and from this one can derive a 'chemical shift' which is characteristic of the environment of the nucleus. For $^{1}H$ and $^{13}C$, this chemical shift $\delta$(ppm), is measured relative to tetramethylsilane (TMS). In $^{1}H$ nmr the area under a peak is proportional to the number of protons having the same chemical shift.

Nuclei possessing a magnetic moment which are close to one another in a molecule interact to produce a mutual coupling which may lead to the spin-spin splitting of the signals at the relevant chemical shifts. The pattern of coupling and the size of the coupling constant, $J$(Hz), contain substantial structural information. For coupling between two nuclei, A and B, $J_{AB} = J_{BA}$. The structural information from $^{1}H$ nmr spectra is considerably amplified and extended by $^{13}C$ nmr spectroscopy. One distinguishes between 'broad-band decoupled' $^{13}C$ spectra, which give only the chemical shifts of the $^{13}C$ atoms, and 'off-resonance' spectra, which retain residual couplings from $^{1}H$ nuclei bonded directly to the $^{13}C$ atoms.

**Ultraviolet and visible spectroscopy** (uv/vis): Band positions and intensities for elecronic transitions supply information concerning the type and extent of conjugated $\pi$ systems.

The determination of the structure of an unknown compound rests on the amalgamation of structural elements derived from spectroscopic data into the most probable combination.

The flow diagram below summarises the procedure.

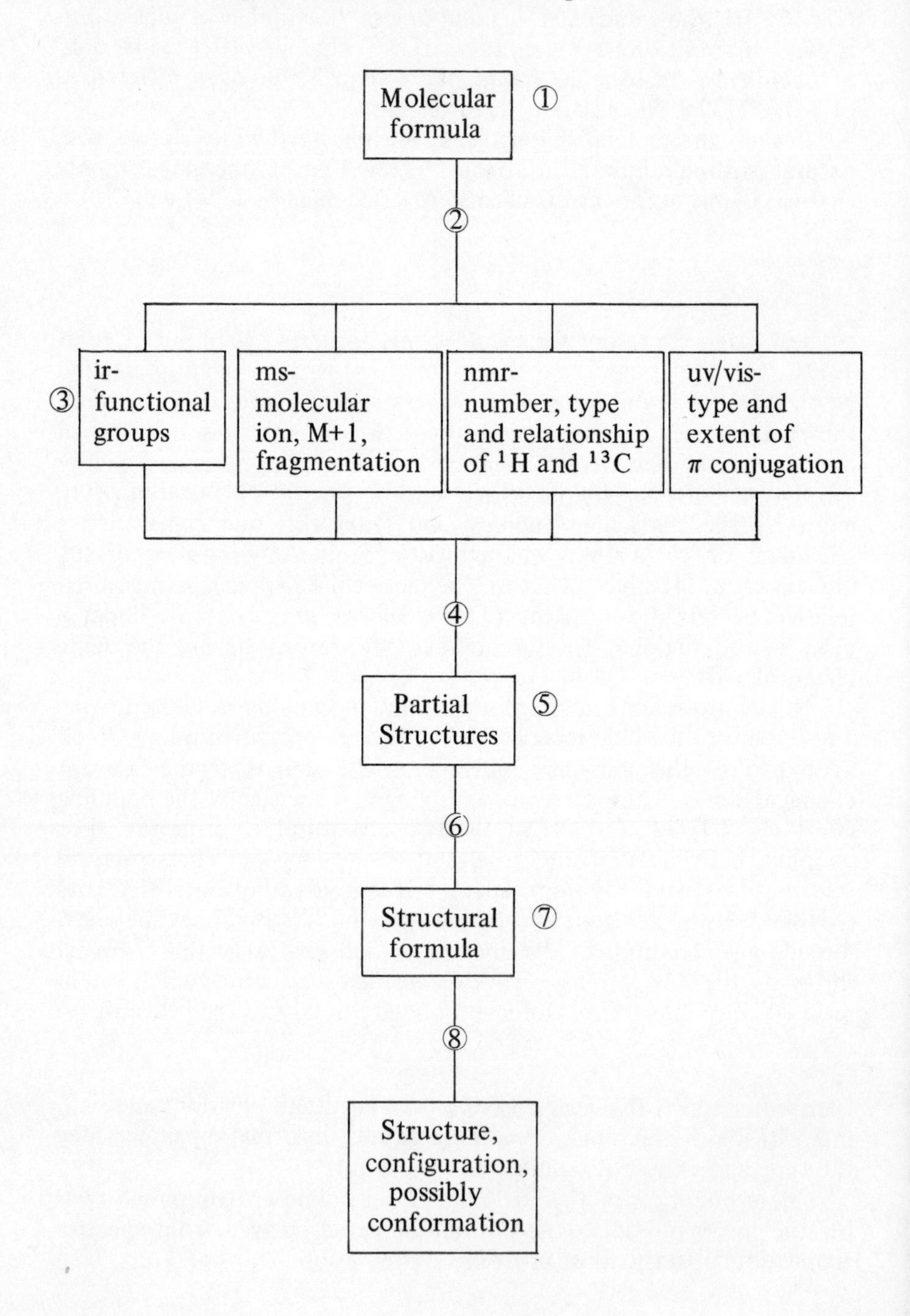

① Elucidation of a structure generally starts with the *molecular formula*. This may be determined from the *empirical formula* from combustion analysis plus the molecular weight. The latter may be obtained either from the molecular ion ($M^{+\cdot}$) in the mass spectrum or by other techniques.
Once the molecular formula is known, calculate $F$, the number of double-bond equivalents (double bonds + rings):

$$F = 1 + \frac{\Sigma(\text{no. of atoms of a particular element}) \times (\text{formal valency} - 2)}{2}$$

In the simpler case of compounds $C_xH_yO_z$, this formula reduces to:

$$F = 1 + \left(x - \frac{y}{2}.\right)$$

In conjunction with the uv/vis spectrum, $F$ provides information on the number of double bonds and rings.

② From the spectra deduce probable structural elements; check whether they are also detectable in the other spectra. In addition, note which functional groups are absent, especially from the ir.

③ ir: analyse particularly the regions 4000–3100, 3100–2700, 1800–1650 $cm^{-1}$
ms: the peak of highest $m/z$ is often, but not always, $M^{+\cdot}$. Typical fragmentation patterns are known for many classes of compound.
nmr: tables of chemical shifts and coupling patterns can often lead to unambiguous partial structures.
uv/vis: tables can lead to the identification of many chromophore structures.

④ Keep a running check between the structural elements determined according to ③ and the molecular formula in order to decide
– how many atoms and groups are still missing
– whether the sum of the partial structures exceeds the molecular formula.

⑤ If a number of partial structures or structural elements are compatible with the molecular formula, consult additional analytical data e.g. fingerprint region in ir, double resonance in nmr.

⑥ Combine the partial structures, taking into account such factors as symmetry in the nmr spectrum, probable fragmentation in ms, etc. If it seems likely that the substance is known, consult tables of melting points, refractive indices, etc., which may lead to a rapid unambiguous confirmation of the proposed structure.

⑦ The final structural proposal must be evaluated in terms of the expected spectral data (tables, additivity rules, spectra of model compounds). All the observed spectroscopic characteristics must be plausibly explicable. If the results do not lead to an unambiguous structure, one or more chemical reactions may lead to a compound whose structure may be determined uniquely by spectroscopic means.

⑧ Configurations and, where appropriate, conformations are best deduced separately from the analytical data.

Deduce the structure of the compound which gives the following data.

Elementary analysis
C 73.17% H 7.32%

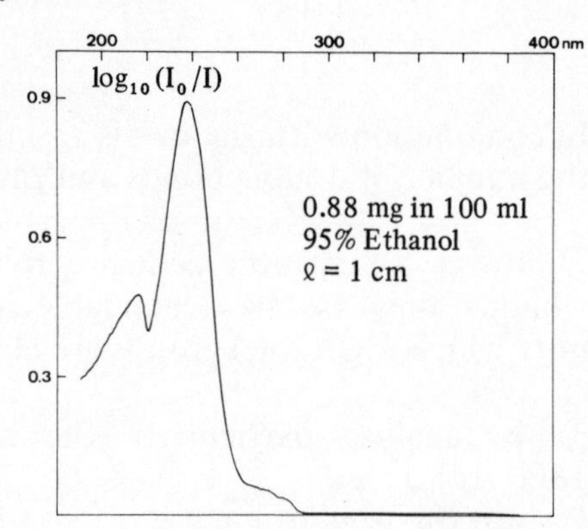


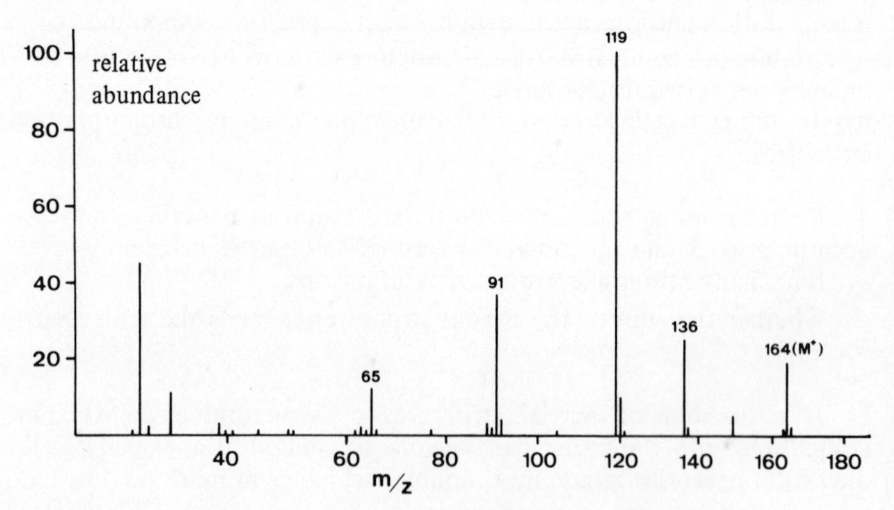

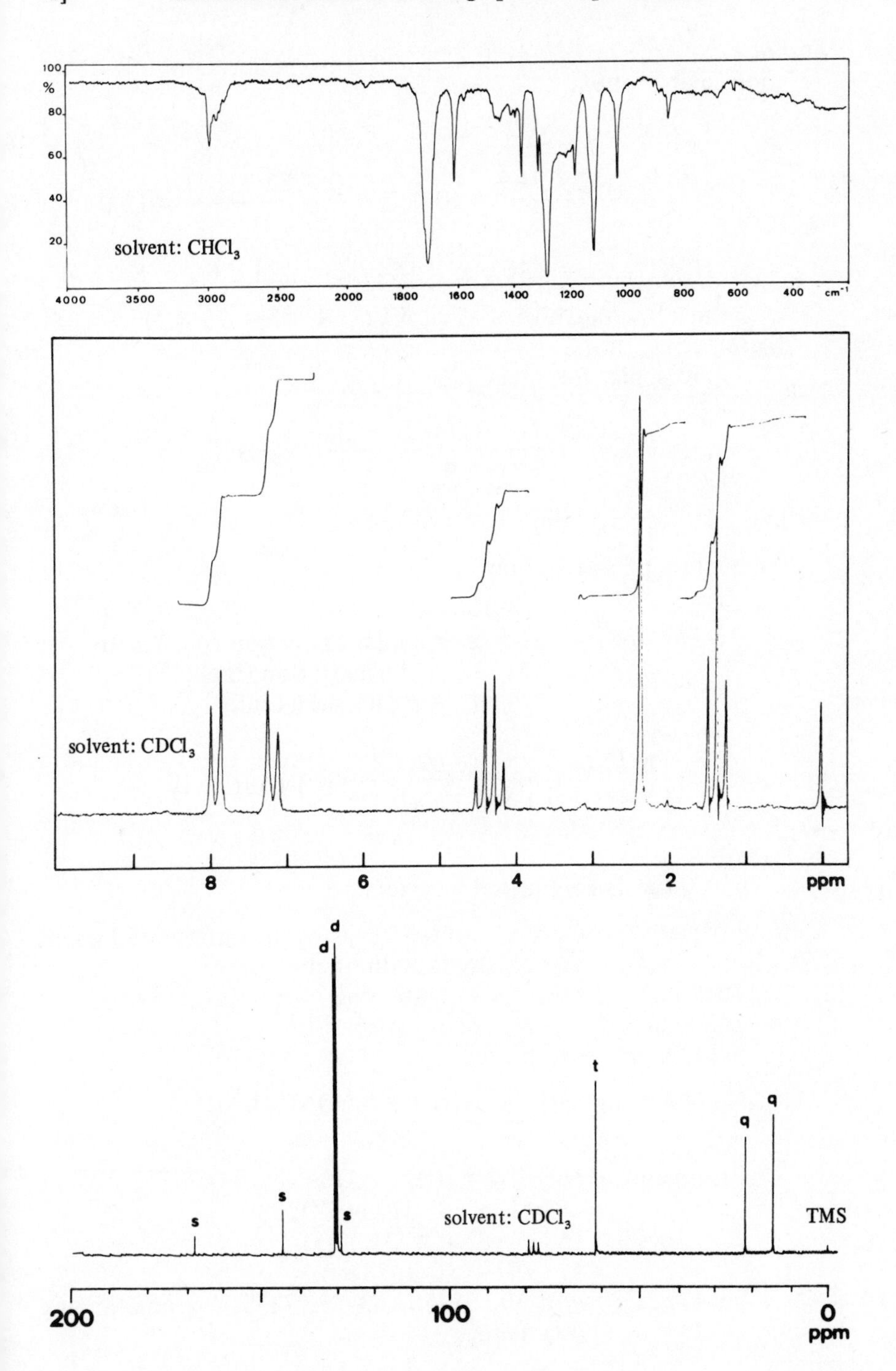
100
%
80
60
40
20
solvent: $CHCl_3$
4000
3500
3000
2500
2000
1800
1600
1400
1200
1000
800
600
400
cm⁻¹
solvent: $CDCl_3$
8
6
4
2
ppm
d
d
t
q
q
s
s
s
solvent: $CDCl_3$
TMS
200
100
0
ppm

*Solution:*

– molecular formula:

C 73.17% H 7.32%; ms: $M^{+\cdot} = 164$

$$C = \frac{0.7317 \times 164}{12} = 10.0 \quad H = \frac{0.0732 \times 164}{1} = 12.0$$

$C_{10}H_{12} \ldots = 132 + x = 164 \therefore x = 32$ i.e. $O_2$ or S

ms: S would give M+2 *ca.* 4% ($^{32}S:^{34}S = 95:4.2$)

therefore $O_2$, not S.

molecular formula $C_{10}H_{12}O_2$

$$F = 1 + \frac{10(4-2) + 12(1-2) + 2(2-2)}{2} = 5$$

– uv: $\lambda_{max} = 237$ nm

Calculation of $\epsilon$ at 237 nm

$$\epsilon cl = \log \frac{I_0}{I}$$

$\epsilon$ = molar extinction coefficient
$c$ = concentration (mol $\ell^{-1}$)
$l$ = path length (cm)

$$\text{therefore } \epsilon = \frac{\log I_0/I}{cl} = \frac{0.89}{1 \times 5.37 \times 10^{-5}} = 1.6586 \times 10^4$$

$\epsilon$ suggests conjugated $\pi$ system.

– ir: 1705 $cm^{-1}$ C = O stretch (aryl or $\alpha\beta$ unsaturated ester, aryl aldehyde, saturated ketone)
1610 $cm^{-1}$ possible aromatic ring
1280 $cm^{-1}$ –C–O–stretch
∿3000 $cm^{-1}$ CH stretch

No bands for –OH, $-CO_2H$(–CN, $-CONH_2$).

– ms: (M+1) *ca.* 10% of $M^{+\cdot}$ ∴9–10 C

Fragments
164→149 ($CH_3$, 15)
164→136 ($C_2H_4$ or CO, 28)
164→119 ($OC_2H_5$, 45)
119→91 (CO, 28)

Weaker: *m/z* 39, 51–53, 63–65, 77–8 – aromatic ring fragments.

– nmr

$^1H$: Intensities: 40:40:40:60:60: → 2:2:2:3:3
$\Sigma 12$, consistent with molecular formula.

Chemical shifts: $\delta =$ 1.35 (t, J 7Hz, 3H)
2.40 (s, 3H)
4.32 (q, J 7Hz, 2H)
7.18 (d?, J 8Hz?, 2H)
7.9 (d?, J 8Hz?, 2H)

– signals at 1.35 and 4.32 are coupled together, with pattern for $-CH_2CH_3$ and chemical shift of $CH_2$ suggesting $-OCH_2CH_3$
– signals at 7.18 and 7.9 are in aromatic region and coupling pattern is typical AA′ BB′ for 1,4-disubstituted benzene ring.

$^{13}C$: broad band decoupled:

| *chemical shift* | *multiplicity* (from 'off resonance' spectrum) | |
|---|---|---|
| 14.34 | q | $(CH_3)$ |
| 21.55 | q | $(CH_3)$ |
| 60.61 | t | $(CH_2)$ |
| 127.73 | s | $(-C\leqslant)$ |
| 128.84 | d | (CH) |
| 129.40 | d | (CH) |
| 143.15 | s | $(-C\leqslant)$ |
| 166.40 | s | $(-C\leqslant)$ |

A least 8 C-atoms. From molecular formula, 2 C-atoms must be duplicated. → 1,4-disubstituted benzene?

– partial structures:

$CH_3$, –CO–O–R, $-OCH_2CH_3$,

– proposed structure:

$CH_3-C_6H_4-C(=O)-OCH_2CH_3$

This structure accounts for:

$F = 5$: 4 double bonds + 1 ring

$^1H$ coupling: $-CH_2$ $-CH_3$; $^3J_{H-H}$ = 7Hz (e.g. in ethyl acetate)

$^{13}C$ chemical shifts: methyl *p*-methylbenzoate has the following signals

21.4 (Ar-$CH_3$); 51.6 ($OCH_3$); 127.9, 129.2, 129.7, 143.4 (Ar); 166.8 (CO).

ms: $M^{+\cdot}$ contains all the elements found in the fragments and by other spectroscopic methods.
*m/z* 136 arises from the rearrangement reaction

Ar—C(=O)—O—$CH_2$—$CH_2$—H, +·


## BIBLIOGRAPHY

J. T. Clerc, E. Pretsch, and J. Seibl, *Structural Analysis of Organic Compounds,* Elsevier, 1981.

T. S. Ma and R. F. Lang, *Quantitative Analysis of Organic Mixtures,* Part 1, Wiley, New York, 1979.

D. H. Williams and I. Fleming, *Spectroscopic Methods in Organic Chemistry,* 3rd edition, McGraw Hill, London, 1980.

R. M. Silverstein, G. C. Bassler, G. Clayton & T. C. Morrill, *Spectrometric Identification of Organic Compounds,* 4th ed., Wiley, New York, 1981.

J. Lambert, H. F. Shurvell, L. Verbit, R. G. Cooks, and G. H. Stout, *Organic Structural Analysis,* MacMillan, New York, 1976.

P. L. Fuchs and C. H. Bunnell, *Carbon-13 NMR Based Organic Spectral Problems,* Wiley, New York, 1979.

### Infrared Spectroscopy

L. Bellamy, *The Infrared Spectra of Complex Molecules,* 3rd ed., Halsted Press, New York, 1975.

D. Dolphin & A. Wick, *Tabulation of Infrared Spectral Data,* Wiley Interscience, New York, 1977.

K. Nakanishi & P. H. Solomon, *Infrared Absorption Spectroscopy,* 2nd edition, Holden Day, 1976.

G. Socrates, *Infrared Characteristic Group Frequencies,* Wiley, New York, 1980.

### NMR Spectroscopy

R. S. Abraham & P. Loftus, *Proton and Carbon-13 NMR Spectroscopy,* Heyden, Philadelphia, 1978.

E. W. Becker, *High Resolution NMR,* 2nd edition, Academic Press, 1981.

J. W. Emsley, J. Feeney, & L. H. Sutcliffe, *High Resolution Nuclear Magnetic Resonance Spectroscopy*, Pergamon, Oxford, 1966.

L. M. Jackman & S. Sternhell, *Applications of Nuclear Magnetic Resonance Spectroscopy in Organic Chemistry*, 2nd ed., Pergamon, Oxford, 1969.

G. C. Levy & G. L. Nelson, *Carbon-13 Nuclear Magnetic Resonance for Organic Chemists*, Wiley, New York, 1972.

**Mass Spectrometry**

F. W. McLafferty, *Interpretation of Mass Spectra*, 3rd edition, Wiley, 1980.

D. H. Williams and I. Howe, *Principles of Organic Mass Spectrometry*, McGraw Hill, 1972.

C. Merritt & C. N. McEwen, *Mass Spectrometry*, Part A, Marcel Dekker, New York, 1979.

**Ultraviolet/visible Spectroscopy**

*Spectroscopy in Organic Chemistry*, Arnold, London, 1967.

C. N. R. Rao, *Ultraviolet and Visible Spectroscopy*, 2nd ed., Butterworth, London, 1967.

CHAPTER 7

# Use of the Chemical Literature

One needs the information contained in the chemical literature when one wishes to ascertain whether certain compounds are known and what their structures and properties are, to prepare or update a review of the state of knowledge in an area of the subject, to avoid duplication of results, to support or verify theories and ideas using published data, or to advance scientific research.

The following introduction, in which a basic approach to the chemical literature will be described, will not only give a brief description of the literature, but will also discuss aspects of typical problems and their solution.

Each year some 300 000 papers, 50 000 patents, and a large number of books are published in the general area of chemistry. A well-defined approach and objective are therefore of advantage.

## A. SEARCH FOR A PARTICULAR COMPOUND BY NAME OR FORMULA

As a first step, it may be worth consulting chemical dictionaries or handbooks to see what synonyms exist!

1. Beilstein **'Handbuch der organischen Chemie'**. (In German) This consists of:

| | | |
|---|---|---|
| Hauptwerk (main edition) | covers literature to 1910. | |
| | Acyclic compounds | Vols 1-4 |
| | Alicyclic compounds | Vols 5-16 |
| | Heterocyclic compounds | Vols 17-27 |
| | Index ('Register') | Vols 28-29 |
| 1$^{e}$ Ergänzungswerk (E I) (1st supp.) | covers literature | 1910–1919 |
| 2$^{e}$ Ergänzungswerk (E II) (2nd supp.) | covers literature | 1920–1929 |
| 3$^{e}$ Ergänzungswerk (E III) (3rd supp.) | covers literature | 1930–1949 |
| 4$^{e}$ Ergänzungswerk (E IV) (4th supp.) | covers literature | 1950–1959 |

Each volume contains a Subject Index, and from E III a Formula Index as well. Complete General Subject Indexes have so far appeared only for the main work and the first two supplements. Of the new Collected Index (for H, E I, E II, E III, and E IV), covering the literature to 1959, so far the Formula and Subject Indexes to volumes 1, 2–3, and 17–18 have appeared.

The main edition is divided into volumes by subject area, and the volume numbers are retained in all the supplements.

In addition, the compounds themselves are further divided into 4877 classification numbers ('Systemnummer') throughout, which eases the problems of locating the relevant literature.

*Suggested strategy:*
– Consult the cumulated formula and/or subject index to the 2nd supplement (E II).
– Note volume and page number.
– As this index also gives volume and page references to the main edition and 1st supplement, the compound and its 'Systemnummer' can be located rapidly.
– Because the classification and volume numbers are retained throughout, once a compound has been located in E II, later references in E III and E IV are easily found (provided that the relevant volume has been published!).
– *Best route:* Consult the General Subject or General Formula Index to the 2nd supplement, or the volumes of the new Collective Index as they appear.

An example of how this works in practice is provided later in this section.

2. **Chemical Abstracts** (founded 1907)

*Chemical Abstracts* appears weekly and contains abstracts of all sorts of publications (books, conference reports, governmental research reports, papers, patents, etc.) over the whole area of chemistry.

(a) Each weekly issue contains:
– numbered abstracts (quoting the references to the original literature)
– a keyword index (based on words selected from the title, abstract, and original article) giving the abstract number.
– a numerical patent index (alphabetically arranged by country)
– a patent concordance (which collects the various "national" numbers of identical patents and gives the orginal abstract number).
– an author index.

(b) For each volume (annual originally, two per year since 1962) an index is provided which replaces the weekly indexes, providing:
– a Formula Index (arranged to give carbon compounds containing C and H in ascending numbers of C, H; all other elements strictly alphabetically by element symbols).
– a Subject Index (more comprehensive than the keyword indexes – give brief title and abstract number).
– an Author Index.
– a Patent Index (and a patent concordance).

The use of computers for indexing has produced changes since Vol. 76 (1972), the main one being:

– separation of the Subject Index into Chemical Substance Index and General Subject Index.
– modification of the system of nomenclature used.

The new system of nomenclature is valid for the period of the 9th Collective Index (Vols 76–85, 1972–1976) and for that of the 10th Collective Index (Vols 86–95, 1977–1981). It is comprehensively described in the CHEMICAL SUBSTANCE NAME SELECTION MANUAL (Vols I and II) (Chemical Abstracts Service, 1978).

For example, 'dimethylsulfoxide' is classified under 'Methane-, sulfinylbis-', and 'diethyl ether' under 'Ethane-, 1,1-oxybis-'.

If a compound or concept cannot be located in the (incomplete) Formula Index, it is strongly recommended that one next consults the latest INDEX GUIDE SUPPLEMENT 1977–1980 CUMULATIVE and then the INDEX GUIDE 1977. A revised, comprehensive Index Guide, covering the period of the 10th Collective Index, appears with the Index to Vol. 95. Formula, Subject, and General Subject Indexes, without the Index Guide, do not lead to a complete coverage of the literature, as many compounds are registered not under their IUPAC names but under less common names.

(c) The Collective Index, covering a 10-year period, consists of Formula, Subject, Author, and Patent Indexes.

To assist the search for ring systems, in 1980 the Ring Index was replaced by the PARENT COMPOUND HANDBOOK. This consists of a Parent Compound File, a catalogue of structure diagrams with, among other things, their Chemical Abstracts names, as well as an Index of Parent Compounds, with which one can identify the structure of a ring system whose name is known. Supplements will appear until 1983.

**3. Chemiches Zentralblatt** (1830–1969) (in German)

*Chemisches Zentralblatt* consists of abstracts accessible through the annual indexes via their index numbers. The indexes are divided, much as in Chemical Abstracts, into Formula, Subject, Author, and Patent Indexes. In addition, there are quinquennial collective indexes for the period 1897–1939.

It is worth using *Chemisches Zentralblatt*
– for items from the Russian and Eastern European literature
– if the original literature is inaccessible (the abstracts are more extensive than those of *Chemical Abstracts*)
– for references before 1907.

*Example of a literature search for a particular compound by name and formula.*

Problem 1: for fulvene ( 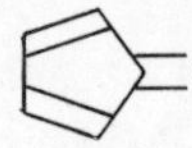 $C_6H_6$), find its syntheses and properties, and the first reference to it.

| Search | Information found |
|---|---|
| *Beilstein:* under "Fulven" in subject index ("Generalsachregister") to E II, Vol. 28 (1st part) p. 1263 | – Fulven: **5**, 280, I 144, which means: Vol. 5 of Main edition ("Hauptwerk"), p. 280, and E I Vol. 5, p.144 (remember volume numbers are retained) |
| *or* | |
| under $C_6H_6$ in the formula index ("Generalformelregister") to E II, Vol. 29 (1st part) p. 161 | – Fulven: **5**, 280, I 144. |
| Then: | |
| Main edition, Vol. 5, p. 280 | –System number ("Systemnummer") **465**!<br>– Thiele, B, **33**, 667 (1900) (B is *Berichte der deutschen chemischen Gesellschaft*) |
| E I, Vol. 5, p. 144 | – a correction to the previous entry |
| E II, | – no futher references |
| E III, Vol. 5, System number 465 (on p. 650) | – 7 more references. |
| This gives the literature to 1949 | |

For the literature 1949–1971, use *Chemical Abstracts*

| Search | Information found |
|---|---|
| – *5th Collective Index* (Vols 41-50, 1947-1956) | |
| Subject Index | – 51 references (with contents indicated by keywords e.g. derivs. **43**: 3796c; **44**: 4430f,.. in Diels-Alder reaction, **50**, 633h . . . electron distribution of, **47**: 401 e. |

| | |
|---|---|
| Formula Index | – 29 references. e.g. Fulvene **42**: 4808h, **43**, 10e, 52460c, P6644a; **44**: 3315h, 4331d, . . . |
| – *6th Collective Index* (Vols 51-55, 1957-1961) | |
| Subject Index | – 43 references |
| Formula Index | – 19 references. |
| – *7th Collective Index* (Vols 56-65, 1962-1966) | |
| Subject Index | – 83 references. |
| Formula Index | – "see Fulvene". |
| – *8th Collective Index* (Vols 66-75, 1967-1971) | |
| Subject Index | – 85 references |
| Formula Index | – 48 references |
| Also: Index Guide (under Fulvene) | – Fulvene |
| *however,* for | |
| Fulvene-, 6-amino- | – see 2,4-cyclopentadiene, methylamine* |
| Fulvene, 6-tert-butyl | – see 1,3-Cyclopentadiene, 5-neopentylidene* |
| Fulvene, 6-phenyl- | – see Methane, 2,4-cyclopentadien-l-ylidenephenyl-* |
| – *9th Collective Index* (vols 76-85, 1972-76) | |
| Index Guide (under Fulvene) | – see 1,3-Cyclopentadiene, 5-methylene [497-20-1]** |
| Subject Index | – 83 references |
| Formula Index. | – 1 reference (!) |

For the literature 1977--1979, consult Vols 86–91.

| Search | Information found |
|---|---|
| – Index Guide Supplement 1977–1979 Cumulative Fulvene | – no entry |
| – Index Guide 1977 Fulvene | – see 1,3-Cyclopentadiene, 5-methylene [497-20-1]** |
| – *Vol. 86 (1977):* | |
| Chemical Subject Index | – 13 references |
| Formula Index | – 13 references |
| – *Vol. 87 (1977):* | |
| Chemical Substance Index | – 11 references |
| Formula Index | – 11 references |

| | |
|---|---|
| – *Vol. 88 (1978):* | |
| Chemical Substance Index | – 11 references |
| Formula Index | – 10 references, and the note: "For general derivatives see Chemical Substance Index" |
| – *Vol. 89 (1978):* | |
| Chemical Substance Index | – 5 references |
| Formula Index | – 5 references, and the note: "For general derivatives see Chemical Substance Index" |
| – *Vol. 90 (1979)* | |
| Chemical Substance Index | – 9 references |
| Formula Index | – 8 references, and the note: "For general derivatives see Chemical Substance Index" |
| – *Vol. 91 (1979):* | |
| Chemical Substance Index | – 6 references |
| Formula Index | – 6 references and the note: "For general derivatives see Chemical Substance Index" |
| – *Vol. 92 (1980):* | |
| Chemical Substance Index | – 5 references |
| Formula Index | – 4 references |
| – *Vol. 93 (1980):* | |
| Chemical Substances Index | – 9 references |
| Formula Index | – 8 references |

For any later volumes for which the annual indexes have not yet appeared, a search of the weekly keyword indexes under 1,3-Cyclopentadiene, 5-methylene- will yield many of the relevant references.

*(These last three examples illustrate the existence of widely differing nomenclature for closely related structures. Without the Index Guide, neither Subject nor Formula Index would yield any trace of them!!)

**In Chemical Abstracts, from the 8th Collective Volume onwards, each chemical substance is uniquely characterised by a *Registry Number,* which is independent of nomenclature and its possible variations. Once one has locacted the Registry Number, for example in the Chemical Substance Index, references to the compound can rapidly be distinguished from those for its isomers in the Formula Index. It is particularly useful that, for example, configurational isomers [(*R*), (*S*)] and racemic forms [(*R,S*)] each have their own Registry Number.

## B SEARCH FOR PARTICULAR REACTIONS
(or classes of reaction)

### 1. In review indexes.

(a) *Index of Reviews in Organic Chemistry,*
by J. A. Lewis, published by The Chemical Society, London.
There are five volumes of this work:

| | | |
|---|---|---|
| Cumulative Issue | 1971 | "comprehensive coverage" of the literature 1960–1970. |
| Supplement | 1972 | 1500 references 1971–72 |
| Supplement | 1973 | 1500 references 1972–73 |
| 2nd Cumulative Issue | 1976 | *ca.* 8000 references, comprising the 1972 and 1973 Supplements plus reviews published 1973–76. |
| Supplement | 1979 | *ca.* 3500 references 1976–78. |

Each volume is arranged alphabetically, with each letter divided into three sections:

Section 1: references to articles about particular compounds or classes of compounds.

Section 2: classification by named reactions.

Section 3: references to chemical processes or phenomena, physico-chemical methods, etc.

(b) *Index to Scientific Reviews* (since 1975), published by The Institute of Scientific Information, Philadelphia.

Appears half-yearly and records more than 16 000 reviews in the whole field of science.

### 2. In review series.

(a) W. Theilheimer *Synthetic Methods of Organic Chemistry.*

Contains short descriptions of individual reactions with experimental details. A good starting-point for literature searches for particular types of reactions. The system of division of volumes by reaction type is explained in Vol. 2.

Each volume has an index of methods and classes of compound. Reagents are quoted separately. Once one has grasped the method of classification the systematic reviews and supplementary references become useful. There is a cumulated index in every fifth volume.

(b) *Organic Reactions* (founded 1942 (Vol. 1) by R. Adams).

In each volume particular types of reaction which are of general application are discussed with the emphasis on preparative aspects. There are references to the original literature. Individual chapters contain comprehensive tabulated compilations.

(c) *Organic Syntheses* (founded 1921 (Vol. 1) by R. Adams).
Volumes appear annually and contain tested "recipes" for syntheses of general interest. Five Collected volumes have also appeared, each containing the corrected material from 10 of the annual volumes.

(d) Houben-Weyl *Methoden der Organischen Chemie.*
In fifteen volumes (some subdivided). They deal with matters such as: general laboratory practice, analytical methods, physical methods, etc., as well as classes of substances e.g. alkanes and cycloalkanes, O, N, P, S, Se, Te, and halogen compounds, organometallics, peptides. Emphasis is placed on preparative aspects of the various classes of compounds.

(e) A. Weissberger *Technique of Organic Chemistry.*
Fourteen volumes covering physicochemical methods, separation, analytical techniques, reaction mechanisms, and organic photochemistry. Revised second or third editions of individual volumes appear periodically.

(f) *Annual Reports of The Chemistry Society,* London (since 1980, The Royal Society of Chemistry).
Issued since 1904, this series presents annually in concise but readable form the most significant results in general, physical, inorganic, and organic chemistry.

(g) *Specialist Periodical Reports* of the Chemical Society, London (since 1980, The Royal Society of Chemistry).
In view of the flood of publications in an ever-increasing number of specialised areas, the appearance of these comprehensive surveys of selected subject areas (32 of them) every few years is extremely welcome.

**3. In monographs**

There are so many monographs covering such a vast range of subjects, it would be best to consult a specialist (or fellow student!) for specific purposes.

## C SEARCH (1) FOR A PARTICULAR SCIENTIFIC PROBLEM, OR (2) FOR AUTHORS PUBLISHING IN A PARTICULAR AREA OF THE SUBJECT

*Science Citation Index* (published by the Institute for Scientific Information Philadelphia).

The purpose of the *Science Citation Index* is to enable one to locate all the authors publishing on a particular problem or in a particular area of the subject. To attain this goal it is only necessary to locate the relevant papers (in the Source Index – an alphabetical list by *author* of all papers published, with their respective titles and references). In order to find other relevant papers one searches the Citation Index, a list (by author) of other writers who have quoted the original paper (only authors' names given, without title or reference). Use of these two lists gives one a comprehensive literature search, but without any abstracts. A similar search would be possible without knowledge of any authors' names if one used a key-word index. In the present work this catalogue is based on permutations of significant concepts from the titles of the papers cited (Permuterm Subject Index).

As the declared aim of the publishers is to have the complete chemical literature for any given year indexed shortly after the end of that year this work provides a rapid means of searching the more recent literature.

The following annual volumes appear:

Source Index
Citation Index
Permuterm Subject Index
Index Guide.

There are also quarterly Source and Citation Indexes.

*Examples of literature searches on particular problems:*

*Problem 2:* Up-to-date organic chemistry textbooks discuss the Cahn-Ingold-Prelog Sequence Rules. Find the original literature references.

| Search | Information found |
|---|---|
| *Science Citation Index,* – Citation Index (e.g. 1975) under Cahn, R. S. (initials from textbook!) | 18 publications by R. S. Cahn, cited (in 1975) in between 1 and 46 papers. |
| To use this information to find the desired references, proceed as follows: | |
| (a) Look up the 18 cited papers. | *Angew. Chem. Int. Ed.* **5**, 385, (1966) *Angew. Chem.* **78**, 413, (1966) |
| (b) Source Index (1975) under each cited paper | Titles are given, and on this basis decide which are likely to contain the desired topic. Then look up the original papers. |

The Citation Index is based on the name of the first author for papers with more than one author. Therefore, in this case, the original papers would have been found under Prelog, V., only if V. Prelog had cited them during 1975.

To complete a literature search on the most recent work in this area (i.e. to answer the question: What is the most recent published material connected with the Sequence Rules?), look up in the *Citation Index* for 1976–81 all the first authors (including Cahn, R. S.) who cited the relevant papers in 1975–80.

*Problem 3:* Rotational barriers in molecules

| Search | Information found |
|---|---|
| *Science Citation Index,* | |
| – Subject Index **7**, Column 4033 (1969), under Barrier-Rotation | – Beaudet R. A. Bemey C. V. etc. (17 authors) |
| – Source Index **5**, 666 (1969) under Beaudet R. A. etc. | – *J. Chem. Phys.* **50**, 2002 (1969) Microwave Spectra of substituted propenes. Barrier to internal rotation of trans-l-Bromopropene |
| etc. | |

The Subject Index of *Chemical Abstracts* (January-June, 1969) contains no entries under 'Rotation barrier', 'Internal rotation', nor 'Barrier, rotation'.

*Problem 4:* DNA-polymerase

| Search | Information found |
|---|---|
| *Science Citation Index,* – Subject Index **8**, 13033 (1969) under DNA-polymerase | – Atkinson M. R. Bedson H. S. etc. [50 names, including *Nature* (anonymous article)]. |
| – Source Index **5**, 425 (1969) under Atkinson M. R. | – *Fed. Proc.* **28**, 347 (1969) Active Centre of DNA-Polymerase see also: Kelly R. B., *Nature* **224**, 495 (1969) etc. (6 further references) |
| etc. under (Anon.) *Nature* | – *Nature* **224**, 1151 (1969) What makes DNA duplicate? |
| etc. | |

The Subject Index of *Chemical Abstracts* (January-June 1969) contain references under neither 'DNA-polymerase' nor 'Polymerase, DNA'.

## D SEARCH FOR SPECTROSCOPIC DATA ON ORGANIC COMPOUNDS

The indentification of organic compounds is becoming increasingly dependent on instrumental methods of analysis. The use of these methods is substantially aided by monographs and collections of data which appear regularly.

The following list indicates some of the current data collections on NMR, IR, UV/visible, and mass spectroscopy.

*NMR Spectroscopy*

| | |
|---|---|
| $^{1}H$ – NMR Spectra Catalog (from 1966) | Sadtler Research Laboratories, Philadelphia. |
| – NMR Spectral Data (40 and 60 MHz from 1959, 100 MHz from 1969) | American Petroleum Institute Research Project, Texas A & M University. |
| – NMR Spectral Data (40 and 60 MHz from 1960, 100 MHz from 1969) | Thermodynamics Research Center Data Project, Texas A & M University. |
| – Formula Index to NMR Literature Data (2 vols.). | M. G. Howell, A. S. Kende, and J. S. Webb, Plenum Press, N. Y., 1965-66. |
| – Handbook of NMR *Spectral Parameters* Vol I-III | W. Brügel, Heyden, London. |
| $^{13}C$ – Carbon-13 *NMR Spectra. A Collection of Assigned, Coded, and Indexed Spectra* | L. F. Johnson & W. C. Janukowski, R. E. Kriegel, New York, 1978. |
| – *Atlas of Carbon-13 NMR Data* (Vols I-III) | E. Breitmaier, G. Haas, & W. Voelter, Heyden, London, 1979. |
| – *Carbon-13 NMR Spectral Data* | W. Bremser, L. Ernst, & B. Franke, Verlag Chemie, Weinheim, 1978-1981. |

*IR Spectroscopy*

| | |
|---|---|
| – *Sadtler Standard Spectra* (from 1962) | Sadtler Research Laboratories, Philadelphia |
| – *DMS Atlas of IR Spectra* | Verlag Chemie, Weinheim, 1972. |

UV/visible Spectroscopy

| | |
|---|---|
| – *Selected UV Spectral Data* (from 1945) | American Petroleum Institute. Project 44. |

| | |
|---|---|
| – *Selected UV Spectral Data* (from 1969) | Thermodynamics Research Centre Data Project |
| – *Organic Electronic Spectral Data* (Vols. I-X, 1946–1968) | H. P. Phillips, H. Feuer, & B. S. Thyagarajan, Interscience, N.Y., 1960–1969. |
| *Mass Spectroscopy* | |
| – *Eight Peak Index of Mass spectra* (2 vols., 1970) | Mass Spectrometry Data Centre, Aldermaston. |
| – *Registry of Mass Spectral Data* | E. Stenhagen, S. Abrahamson, & F. W. McLafferty, Wiley-Interscience, N. Y., 1974. |
| – *Mass Spectra of Compounds of Biological Interest* (2 vols.) | Technical Information Centre (AEC) National Technical Information Service, Springfield, 1974. |
| – *Compilation of Mass Spectral* Data (3 vols.) | A. Cornu & R. Massot, Heyden, 1966–1971. |

## BIBLIOGRAPHY

R. E. Maizell, *How to Find Chemical Information,* Wiley Interscience, New York, 1979.

H. M. Woodburn, *Using the Chemical Literature,* Marcel Dekker, Inc., New York, 1974.

*Use of Chemical Literature,* R. T. Bottle ed., 3rd edition, Butterworths, 1979.

O. Weissbach, *The Beilstein Guide,* Springer Verlag, 1976.

## Standard abbreviations for some of the more common journals

For a complete listing, see Chemical Abstracts Service Source Index.

| | |
|---|---|
| *Acc. Chem. Res.* | *Accounts of Chemical Research* |
| *Acta Chem. Scand.* | *Acta Chemica Scandinavica* |
| *Anal. Chim. Acta* | *Analytica Chimica Acta* |
| *Angew. Chem.* | *Angewandte Chemie (*continuation of *Die Chemie)* |
| *Angew. Chem. Internat. Ed. Eng.* | *Angewandte Chemie, International* Edition in English |
| *Ann. Chim. (Paris)* | *Annales de Chimie* |

| | |
|---|---|
| *Arch. Pharm.* | *Archiv der Pharmazie* |
| *Ber. Bunsenges, Phys. Chem.* | *Berichte der Bunsengesellschaft für Physikalische Chemie (continuation of Z. Elektrochem.)* |
| *Ber. Dtsch. Chem. Ges.* | *Berichte der Deutschen Chemischen Gesellschaft (*since 1947, *Chemische Berichte)* |
| *Biochem. J.* | *The Biochemical Journal* |
| *Biochem. Z.* | *Biochemische Zeitschrift* |
| *Biochim. Biophys. Acta* | *Biochimica Biophysica Acta* |
| *Bull. Chem. Soc. Jpn.* | *Bulletin of the Chemical Society of Japan* |
| *Bull. Soc. Chim. Fr.* | *Bulletin de la Société chimique de France* |
| *Can. J. Chem.* | *Canadian Journal of Chemistry* |
| *C. A.* | *Chemical Abstracts* |
| *Chem. Ber.* | *Chemische Berichte (*continuation of *Ber. Dtsch. Chem. Ges.)* |
| *Chem. Commun.* | *Chemical Communications* (since 1972 *J. C. S. Chem. Commun.*) |
| *Chem. Rev.* | *Chemical Reviews* |
| *Chem. Zentralb.* | *Chemisches Zentralblatt* |
| *C. R. Hebd. Séances Acad. Sci.* | *Comptes rendus des Séances de l'Académie des Sciences, Paris* |
| *Eur. J. Biochem.* | *European Journal of Biochemistry* |
| *Gazz. Chim. Ital.* | *Gazzetta Chimica Italiana* |
| *Helv. Chim. Acta* | *Helvetica Chimica Acta* |
| *Inorg. Chem.* | *Inorganic Chemistry* |
| *J. Am. Chem. Soc.* | *Journal of the American Chemical Society* |
| *J. Biol. Chem.* | *Journal of Biological Chemistry* |
| *J. Chem. Educ.* | *Journal of Chemical Education* |
| *J. Chem. Soc.* | *Journal of the Chemical Society, London,* 1966–1971 in Sections A, B, and C. Since 1972 in five *Transactions*<br>*Dalton Trans.* (Inorganic Chemistry)<br>*Perkin Trans. I* (Organic and bio-organic)<br>*Perkin Trans. II* (Physical Organic)<br>*Faraday Trans. I* (Physical)<br>*Faraday Trans. II* (Chemical physics) |
| *J. Chem. Phys.* | *Journal of Chemical Physics* |

| | |
|---|---|
| *J. Heterocycl. Chem.* | *Journal of Heterocyclic Chemistry* |
| *J. Organomet. Chem.* | *Journal of Organometallic Chemistry* |
| *J. Polym. Sci.* | *Journal of Polymer Science* |
| *J. Prakt. Chem.* | *Journal für praktische Chemie* |
| *Justus Liebigs Ann. Chem.* | *Liebigs Annalen der Chemie* |
| *Kolloid–Z.* | *Kolloid–Zeitschrift* |
| *Monatsh. Chem.* | *Monatshefte für Chemie* |
| *Org. React.* | *Organic Reactions* |
| *Org. Synth.* | *Organic Syntheses* |
| *Org. Synth. Coll. Vol.* | *Organic Syntheses Collective Volume* |
| *Pharm. Acta Helv.* | *Phamaceutica Acta Helvetiae* |
| *Q. Rev.* *Chem. Soc. Rev.* | *Quarterly Reviews (*since 1972, *Chemical Society Reviews)* |
| *Recl. Trav. Chim. Pays-Bas* | *Recueil des Travaux Chimique des Pays-Bas* |
| *Rev. Chim. Mineral.* | *Revue de Chimie Minerale* |
| *Tetrahedron Lett.* | *Tetrahedron Letters* |
| *Theor. Chim. Acta.* | *Theoretica Chimica Acta* |
| *Trans. Faraday. Soc.* | *Transactions of the Faraday Society* |
| *Z. Analyt. Chem.* | *Zeitschrift für Analytische Chemie* |
| *Z. Anorg. Allg. Chem.* | *Zeitschrift für Anorganische und Allgemeine Chemie* |
| *Z. Elektrochem.* | *Zeitschrift für Elektrochemie* |
| *Z. Kristallogr.* | *Zeitschrift für Kristallographie* |
| *Z. Anorg. Allg. Chem. or Leipzig)* | *Zeitschrift für Analytische und* (Frankfurt or Leipzig). |
| *Z. Physiol. Chem.* | *Zeitschrift für Physiologische Chemie* |

CHAPTER 8

# Laboratory Notebooks

An entry in a laboratory notebook constitutes the primary reference to any experiment one has personally carried out. It claims priority in any case of doubt.

All relevant aspects of a conversion should be recorded, together with the order in which steps were carried out. All observations should be noted, in principle even those that at first sight appear unimportant. Only in this way can one ensure that the results of critical experiments can be reproduced.

At the beginning of each experiment record:

- the date.
- structural formulae (abbreviated, if necessary) and all reagents in order of addition.
- molecular formulae and molecular weights, preferably under the relevant structural formulae.
- literature references on the procedure (or on analogous preparations).
- weights (and number of moles) of each compound used.

- list of apparatus (with sketches in unusual cases).
- the purity of all compounds and solvents (which should have been determined!) e.g. "Analar grade", "Freshly distilled from $LiAlH_4$", "filtered through basic alumina, Woelm, activity I", "single spot on t.l.c. ($SiO_2$, hexane)", "pure by n.m.r.", etc.

During the course of the experiment keep running notes on:
- all observations (described exactly).
- the order of individual operations and the time taken over each of them.
- all experiments used to keep a check on the course of the reaction.

At the end of the reaction record without delay:
- the method of working up.
- purification procedures.
- yields and percentage yields.

*Some hints:*
- Use a book with numbered pages.
- A book giving a carbon copy is a valuable safeguard!.
- Write your report on one side of the page only, using the facing page for recording weighings, distillation records, titrations, preliminary tests for separating mixtures, hydrogenation curves, suggested improvements for future use, interpretation of spectra, etc.
- Note reference numbers of spectra in the margin (but keep the spectra themselves in separate files).
- An index makes it easier to retrieve the information.
- For key experiments, write an edited report for later use and file it separately.
- Some research workers also keep a diary to provide a very brief record of what work they have done each day.

25.5.70

p. 43

$C_6H_5$–CH=CH–$C_6H_5$ + $H_2N$–N(phthalimide) $\xrightarrow[CH_2Cl_2]{Pb(OAc)_4}$ trans-2,3-diphenylaziridin-1-yl–N(phthalimide)

$C_{14}H_{12}$ 180 — $C_8H_6N_2O_2$ 162 — $C_{22}H_{16}N_2O_2$ 340

Reference: L. A. Carpino and R. K. Kirkley J. Am. Chem. Soc., **92**, 1784 (1970)

Apparatus: 300 cm 3-neck flask; mechanical stirrer with Teflon blade

Reagents: 40 mmol

6.50 g (40 mmol) N-aminophthalimide (m.p. 199-202°. It solidifies after melting, perhaps because of rearrangement to [phthalhydrazide structure])

36.0 g (0.2 mol, 5-fold excess) trans-stilbene (Fluka, "puriss.")

20.0 g (ca. 40 mmol) lead tetra-acetate (Fluka, "purum", 85-90% moistened with acetic acid)

N-aminophthalimide (6.50 g, 40 mmol) and trans-stilbene (36.0 g, 0.2 mol) were suspended in dichloromethane (100 $cm^3$, distilled over $P_2O_5$) and treated at room temperature and with vigorous stirring over 10 min with lead tetra-acetate (20.0 g, ca. 40 mmol), added in portions. (The reaction mixture turned orange. Virtually no heat was evolved).

The mixture was stirred for a further 30 min, filtered through Celite and evaporated (rotary evaporator, water bath at about 40°C). The residue (a dark brown oil) was immediately chromatographed on Kieselgel (190 g).

p. 44

Dichloromethane eluted excess stilbene, a small amount of an unidentified byproduct

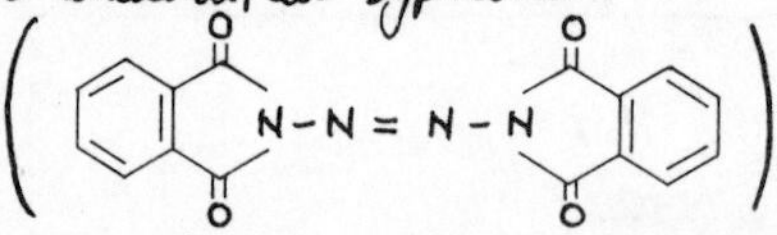

( [phthalimido]N–N=N–N[phthalimido] ) and a yellow crystalline compound (10.0 g, 73.5% calculated as 1-phthalimido-trans-2,3-diphenylaziridine. The product was recrystallised from chloroform-pentane.

Fraction 1 (crystallisation overnight at 0°; dried for 2 h at 0.05 torr/room temp.)
5.48 g (m.p. 177-179°C, lit. (Carpino) m.p. 165°C)
tlc (Kieselgel)

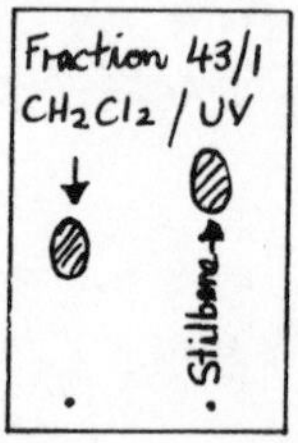

Fraction 43/1
$CH_2Cl_2$ / UV
Stilbene

canary-yellow long needles
IR 43/1 (agrees with lit.!)
Characterisation IR 43/1 (5% $CHCl_3$)
NMR 43/1
MS 43/1 (200°C)
C-H-N microanalysis

IR 43/1
NMR 43/1
MS 43/1
microanalysis

Fraction 2 (ditto)
4.44 g (m.p. 176-177°C, tlc identical with Fraction 1)
IR 43/2 (identical with IR 43/1)

IR 43/2

Yield: 9.92 (73%)

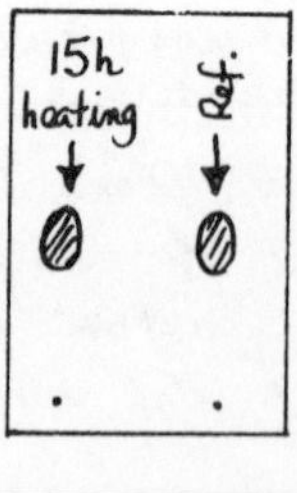

15h heating
Ref.

Thermal stability: A sample was heated at 100°C for 15 h in an open fusion tube. It showed no change (tlc, IR 43/3).

IR 43/3

CHAPTER 9

# Writing a Report

A good clear report is easy to produce if one has a comprehensive description of the work including all relevant data on the starting materials and products as well as all the experimental details in one's laboratory notebook.

The experimental procedure should be described concisely with neat formulae and relevant references.

A report on a preparative experiment should have the following features:

*Title:*
This should include the name of the product, the names of the experimenters, and, where relevant, the date.

e.g. "Isolation of cyclopenten-3-one from. . . ."
"Synthesis of cyclopenten-3-one from. . . .".

*Report:*
This can be arranged in the following sections:

(a) Method: Here the overall transformation carried out in each step of a multi-stage synthesis is described.

e.g. "Pinacol is prepared by the reductive dimerisation of acetone".
"Endo-bicyclo[2,2,1]hept-2-ene-5-carboxylic acid is formed by the cycloaddition (Diels-Alder Reaction) of cyclopentadiene and acrylic acid".

(b) Reaction scheme: This shows the transformation of starting materials to products by means of formulae (configurational or conformational if necessary). Reaction conditions (reagents, solvents, catalysts, temperature, etc.) are indicated in abbreviated form above and below the arrows in the usual way. The molecular formulae and molecular weights can be appended to the relevant structures. All structures should be numbered. In general, diagrams showing mechanisms are presented separately.

(c) Experimental section: The description of the experiments (past tense, third person, passive voice) should be sufficiently detailed to permit the repetition of the reaction without further consultation of the literature. The report should be sufficiently complete for it to be used in preparing a paper for publication. The weights of all compounds (and the number of moles), the purity of starting materials and solvents (see p. 95–6) (these data on substances used repeatedly in a series of experiments can be collected and placed at the beginning of the experimental section, if desired), and all relevant reaction conditions (temperature, time, pressure, etc.) should be quoted, as well as the work-up and purification procedures. One should also provide information about the apparatus used, any peculiarities observed, and simple procedures for following the course of the reaction.

In the text, names of all chemicals should be written out in full; formulae are used only in reaction schemes. On the other hand abbreviated or trivial names (with the structure numbers used in the reaction schemes) make it easier to follow descriptions when long and complex names are involved.

The yield is quoted (*not* the average yield over several preparations) with an indication of purity ("crude", "after recrystallisation", etc.), as well as the literature yield with reference (where relevant).

Finally, the physical data used to characterise the compound should be reported (again with literature references):
m.p., b.p., $n_D$ (with temperature superscript), $R_f$ (with details of t.l.c. system), i.r., u.v., n.m.r., m.s., etc.

(d) Additions and results from other experiments and suggestions for further experiments: This section is necessary if there is no general discussion of the reaction, if variation in any reaction parameter leads to widely different results, if the method of work-up is critical, or if alternative attempts to isolate the product have failed.

*Some typical expressions and abbreviations:*

- bright yellow crystals (10.5 mg, 78% based on *8*)
- tetramethylsuccinic anhydride (33.8 g, 0.217 mol)
- nitrile (1.15 g, 8.5 mmol)
- a solid residue (68.9 g) remained, and was recrystallised from . . . (*ca.* 250 $cm^3$) with charcoal decolorisation
- in absolute ethanol (2 $cm^3$)
- poured onto ice (1.5 kg)
- in sodium hydroxide solution (1 *N*)
- with methyllithium in ether (1.49 M, 16 $cm^3$)

– after addition of hydrochloric acid (18% solution, 400 $cm^3$)
– 3 sec
– 4 min
– 2.5 h
– 50–60°C
– the fraction b.p. 179–195°C/2 torr
– a sample subliming at 110°C/0.005 torr
– in a 100 $cm^3$ round-bottomed flask fitted with reflux condenser and stirrer
– tetramethyllaevulinic acid (*13*) (9.15 g, 53 mmol) was converted into the acid chloride (*14*) by treatment with thionyl chloride (6 $cm^3$)
– oxidation with Fetizon's reagent[4] . . .

EXAMPLE OF A LABORATORY REPORT

1-Phthalimido-trans-2,4-diphenylaziridine

19 July 1975 John Smith

(a) *Method*

trans–Stilbene reacts with N–aminophthalimide in the presence of lead tetra–acetate in a oxidatively induced (2 + 1) cycloaddition to yield 1-phthalimido-trans-2,3-diphenylaziridine([1,2]).

(b) *Scheme*

| $C_6H_5CH{=}CHC_6H_5$ (trans) | $H_2N{-}N$ (phthalimide) | $\xrightarrow[\text{RT}]{Pb(OAc)_4,\ CH_2Cl_2}$ | 1-phthalimido-trans-2,3-diphenylaziridine |
|---|---|---|---|
| $C_{14}H_{12}$ | $C_8H_6N_2O_2$ | | $C_{22}H_{16}N_2O_2$ |
| mw: 180 | 162 | | 340 |

(c) *Experimental*

N-Aminophthalimide(3) (6.50 g, 40 mmol) and trans–stilbene(4) (36.0 g, 200 mmol) were vigorously stirred in dry dichloromethane(5) (100 $cm^3$) in a three-necked 500 $cm^3$ flask fitted with a Teflon-bladed stirrer, a thermometer, and a powder funnel. Lead tetracetate(6) (20.0 g, 40 mmol) was added at room temperature to the suspension over 10 min. After a further 30 min stirring, the mixture was filtered through Celite and concentrated on a rotary evaporator at 40°C. The crude product was at once transferred to a silica gel (190 g) column and the excess stilbene eluted with dichloromethane. A second fraction containing a small amount of an unidentified by-product was then eluted, followed by 1-phthalimido-trans-2,3-diphenylaziridine (10 g).

The product was recrystallised overnight at 0°C from chloroform/pentane and the yellow needles dried for 2 h at 25°C/0.05 torr. The yield of recrystallised material m.p. 177-179°C was 5.48 g. A second crop of crystals (4.44 g), m.p. 176–177°C, was obtained from the mother liquor.

Yield after chromatography: 10.0 g (73.5%)
Yield after crystallisation: 9.92 g (73.0%).

(d) *Physical data*

$R_f$ (t.l.c., silica gel, dichloromethane): 0.44

m.p. 177–179°C (lit.[1] m.p. 165°C)

$\upsilon_{max}$ (chloroform): 1774 (m), 1718 (s) $cm^{-1}$

$\delta(CDCl_3$, 60 MHz): 3.96 and 4.97 (2H, ABq, J 6 Hz), 7.08–7.83 (14 H, m) ppm

m/z 340 ($M^+$, 9%), 194 (100)

Found: C, 77.55; H, 4.79; N, 8.29. Calculated for $C_{22}H_{16}O_2N_2$, C, 77.63; H, 4.74; N, 8.23%.

(e) *Notes*

If the filtrate obtained by filtration of the reaction mixture through Celite is treated with at equal volume of pentane at 0°C, instead of being concentrated, a crystalline precipitate is produced. Recrystallisation from pentane/dichloromethane gives a yield of 55–65%.

Heating of the reaction mixture at 100°C for 15 h did not affect the result (i.r., t.l.c.).

(f) *References*

1 L. A. Carpino and R. K. Kirkley, J. Am. Chem. Soc., 1970, 92, 1784.
2 D. J. Anderson, T. L. Gilchrist, D. C. Horwell, and C. W. Rees, J. Chem. Soc.(C), 1970, 576.
3 Fluka, puriss.
4 Fluka, purum, m.p. 199–202°.
5 Distilled over phosphorous pentoxide.
6 Fluka, purum, 85–90%, moistened with acetic acid.

# CHAPTER 10

# Hints on the Synthesis of Organic Compounds

The gradual depression of the melting point leads one to the conclusion that, far from the 3-indolylacetic acid increasing in purity, the sample was in fact being enriched in the impurities.

condition.[6] As after the third experiment no reasonable cause remained to account for the decomposition as all possible sources of adventitious moisture had now been eliminated, the dream of a pheromone synthesis was buried with a heavy heart and an alternative experiment undertaken.

As the solid could not be filtered off by any of these techniques, it was sedimented in the ultracentrifuge. The aqueous supernatant contained brown oil and white solid bodies. The desired product was believed to be these white bodies, so they were washed twice with water. This treatment, however, revealed that they were only common salt, and that the substance was in fact the brown oil. This error naturally led to severe losses.

It seemed probable that the compound had been produced. The extraction with bicarbonate solution was therefore repeated, using ethyl acetate in place of ether, but again no product was obtained.

evaporated in order to allow the benzalacetophenone to crystallise. This did not, however, occur. An analogous attempt using alcohol failed equally. The only remaining alternative seemed to be

saturated brine and extracted with ether (3 x 100 $cm^3$). After distillation of the ether the residue (ca. 0.5 $cm^3$) was subjected to short path distillation. This procedure produced no result, however (no yield).

hydrolised. The large loss on the first recrystallisation was caused by the use of a long-necked flask.

Of the 2nd substance very little was eluted as the column had run dry overnight for some reason.

yield: 7.025 g (0.047 mol, 31%) (lit. yield 87%).

The poor yield may be ascribed to the use of an unsuitable distillation apparatus.

Preparative organic chemistry is a variable amalgam of science, art, and craft, with the quest for new compounds or conversions as its goal. It is often difficult (and sometimes impossible) to bring about a desired transformation. As a result, between one's original concept and the joy of success lie deserted acres of vain endeavour and disappointment. Things can be very tough!

The 'Hints on Organic Synthesis' given below should not be considered as a sure recipe for success if followed rigidly, nor yet as a comprehensive list of errors. They are rather a contribution, mostly the fruit of bitter experience, to the more effective organisation of research work.

The aim of an organic synthesis is the preparation in one or more stages of a particular organic compound. Alternative routes will always present themselves: a great deal depends on the choice of the best one. What criteria govern this choice?:

- Availability of starting materials and reagents (commercial products? delivery times!? given in *Organic Syntheses*?)
- Contribution to chemical knowledge.
- Dangers (poisonous? explosive? inflammable?).
- Amount of time and labour required.
- Cost.
- Environmental effects.
- Availability of good literature preparations.

Once one has decided on a synthetic route, one needs to sort out the following matters:

**Procedure**: If a literature procedure is available, read it carefully (including footnotes, which in, for example, *Organic Syntheses* may contain vital information). If the preparation is not reported in the literature, it may be possible to modify published procedures (e.g. in *Organicum*) or syntheses of related compounds. As far as possible one should endeavour to understand the reaction pathway. Careful comparison of the literature reports on related conversions will often indicate which reaction parameters are critical, where crucial phases in the reaction lie, and which side-reactions are to be avoided. Creative geniuses can use their knowledge, intuition, and experience to devise experimental procedures which are completely novel with appropriate precautions!

**Plan of Work**: The stages to be considered are: reaction – work-up – purification – characterisation.

Develop alternative solutions and weigh them up against each other.

In order to choose the optimal conditions for working-up and purification, the properties of the product should be known or at least estimated (state of aggregation, solubility, m.p., b.p., reactivity e.g. towards water).

Consider whether the reaction is exothermic or endothemic. With large quantities, exothermic reactions may become too hot! In such cases initial heating will need to be carefully controlled, and arrangements should be made for rapid removal of heat and application of cooling if it should become necessary.

**Planning the use of time**: An estimate of the actual duration of each step in the procedure is an advantage. In particular, attention should be directed to establishing at which stages the process could be interrupted if necessary. Beware of a tendency to underestimate the time taken to work-up.

**Scale**: It is important to estimate this. Choose a scale that makes handling easy. A useful rule of thumb: use enough starting material to give a theoretical yield of 10–20 g. In many cases the preparation will need to be repeated several times. Experience shows: the second preparation usually goes better than the first! Large amounts of flammable solvents are always dangerous.

**Preliminary work**: Check availability of all materials (stores, reagent shelves, orders, etc.). Purify reagents and solvents, and set up apparatus. Check all chemicals required for identity and purity (appearance, smell – but don't inhale, m.p., b.p., $n_D$, n.m.r., g.l.c., t.l.c., etc.). Make sure you know the methods for destruction of excess active reagents (alkali metals, catalysts – pyrophoric when filtered, hydrides, toxic materials), see pp. 118–9.

**Preliminary tests**: It is a good idea to try out all reactions on about 1/10th of the planned scale. For completely new reactions, even smaller scale tests are desirable in order to see if the reaction takes place at all.

**Apparatus**: Heterogeneous mixtures greater than 10–20 $cm^3$ require adequate mechanical stirring to reduce concentration and temperature variations. Magnetic stirrers are appropriate for homogeneous solutions.

Reaction vessels should be of adequate size (i.e. never more than $\frac{2}{3}$–$\frac{3}{4}$ full). The only sealed apparatus to be used are autoclaves and ampules: unwittingly sealed apparatus is a classic cause of accidents! If air or moisture have to be excluded use a drying tube or the valve shown on p. 100.

**Laboratory Notebook**: (See the relevant section, pp. 83–4)

**Reaction**: In order to have a reproducible reaction, one must have well-defined reactions parameters (concentrations; temperature – thermometer in the reation mixture; exclusion of oxygen, moisture, light, etc.; reaction time). If possible, one should monitor the course of the reaction by some simple technique (t.l.c., pH, appearance of a precipitate, spot texts, e.g. KI/starch paper for starting materials or product, spectroscopy, g.l.c.).

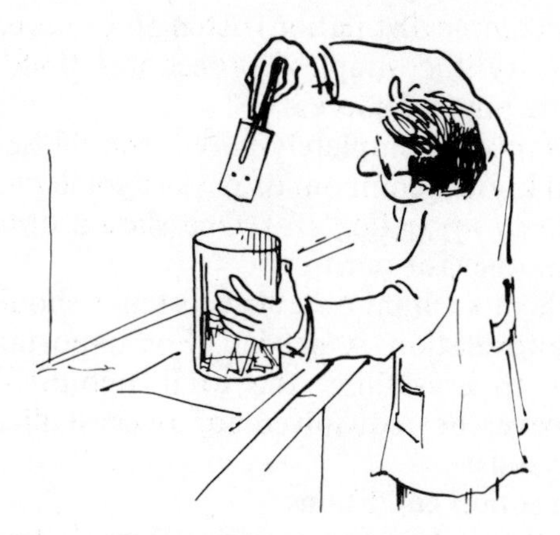

Of course, the 'well-defined reaction parameters' should include an unambiguous termination to the reaction (without uncontrolled further reactions)! This may be achieved by cooling, dilution, removal or decomposition of reagents, etc.

**Work-up and Isolation**: The method of working-up a reaction mixture depends on the relative properties of the product and of any other materials present. One should always consider all the alternatives and choose the most efficient one.

In nine cases out of ten the following standard method will do: pour onto ice, extract with solvent, wash organic layer (neutral/acidic/basic as appropriate), dry, evaporate solvent (rotary evaporator for ether, dichloromethane, pentane, while for products of low b.p. distillation of the solvent via a Vigreux column is often advantageous. *Note:* The more volatile solvents are not completely condensed in a rotary evaporator and are drawn down the water pump to plague water purification plants). Weigh the crude product.

Now ask the *key question:* is this the desired compound? *Yes!; no!;* or more often: *perhaps!!*

Use i.r./n.m.r./t.l.c./g.l.c. to test its purity, and with luck to give an unambiguous answer to the question.

**Purification**: Because many substances are very unstable when impure, a crude product should not be left any longer than necessary. Decomposition can often be delayed by low temperature, storage in the absence of moisture and light, or by the addition of stabilisers.

If the method of purification has not already been established, use only portions of the crude product to try out various techniques. Often decomposition occurs during purification operations.

In practice, the transfer of a purification technique to one of higher capacity (t.l.c. → preparative t.l.c. or column chromatography, bulb distillation → column distillation) often produces complications. Principle: Always try the simplest procedures first, even if they appear to have little hope of success.

Once the substance is completely pure it should be characterised at once. Even well-known compounds require some characterisation (m.p., b.p., i.r., t.l.c., $n_D$, g.l.c., etc). One should obtain one value for identity and another for purity.

A small sample of each pure stable substance should be retained in case further information is needed. For important substances that are difficult to crystallise, the total amount of crystalline material should never be redissolved for re-crystallisation: always save some seed crystals!

**Optimisation of reaction conditions.**

Make a list of all the factors that could influence the course of the reaction (with some idea of how much they might vary).

Then try the reaction under the conditions you judge to be most favourable. If this attempt does not produce any improvement try varying the above factors one at a time to a considerable extent (for optimisation it is essential to vary only one parameter at a time, otherwise it will not be possible to say what caused an improvement or deterioration). Try to use an economical method for estimating the yield.

– Isolate and characterise by-products. One can often glean useful information in this way.

– Results are the consequence of perserverance coupled with ingenuity. Sheer obstinacy seldom leads to success.

## BIBLIOGRAPHY

*Reagents for Organic Synthesis,* Vols. 1–8 (L. F. Fieser and M. Fieser, 1967–1980), Vol. 9. (M. Fieser, R. L. Danheiser & W. Roush, 1981), Wiley Interscience.

I. Harrison & S. Harrison, *Compendium of Organic Synthetic Methods* Vol. 1 (1971), Vol. 2 (1974), Vol. 3 (L. S. Hegedus and L. G. Wade, 1977), Vol. 4 (L. G. Wade, 1980).

*Organic Syntheses*, Vol. 1–60.

H. O. House, *Modern Synthetic Reactions,* 2nd ed., 1972.

*Journal of Synthetic Methods,* Derwent, London.

CHAPTER 11

# Some Tips on Handling Water and Air Sensitive Substances

The manipulation of air or moisture sensitive substances is now a common part of general laboratory practice. The following basic procedures will cover many situations, but will need to be modified considerably for special cases. Between passing a stream of nitrogen through a reaction mixture and working in a dry-box filled with an inert gas containing only a few ppm of oxygen, or on a vacuum line, there are many stages. The simplest aids for handling air or moisture sensitive compounds are characterised by being able to be applied when needed, merely amplifying existing glassware. The following 'Tips' are to be seen as suggestions that can lead to improved technical solutions.

## SIMPLE OPERATIONS UNDER INERT GAS

*Which protective gas to use*

The most common protective gas is nitrogen, with argon as next choice.

*Nitrogen:* This is available in various purities. For many practical applications one needs a quality of 99.995%, although this still contains 10 ppm of oxygen and less than 10 ppm water.

*Argon:* In those cases where $N_2$ can react (e.g. in the preparation of lithium sand, or with complexing transition metals), argon should be used. Again, different qualities are available and for most applications one should use 99.998%, which contains less than 2 ppm of oxygen and less than 3 ppm of water.

Argon has an advantage over nitrogen in that, on account of its greater density, argon-filled vessels can be opened without the need for precautions provided there is no turbulence.

PVC and rubber tubing are permeable to oxygen, so it is advisable to keep passing through a stream of protective gas for the duration of the reaction.

*Secondary purification and drying of protective gases:* Oxygen can be removed using special catalysts e.g. BTS–catalyst (BASF). The commercial catalyst is grey and must first be activated by reduction with $H_2$ at 180–200°C (a few ml of water are formed from each 100 g of catalyst). The active catalyst is dark reddish-brown. Traces of water can most easily be removed using 3Å molecular sieve (p. 123).

*How to prepare the reaction vessel.*

Ideally, the reaction vessel should be heated overnight at 125°C in a drying oven together with any requisite dropping funnels, stirrers, condensers, etc.; the whole apparatus is assembled while still hot and then allowed to cool under protective gas. In some cases it is sufficient to heat the apparatus carefully with a hairdrier while passing protective gas through it.

When passing gas over or through a solution the valve shown in the figure has been found ideal. It allows one to admit the gas and at the same time check its flow. Reaction vessels fitted with such a valve can be opened briefly because a counter-flow of inert gas is formed.

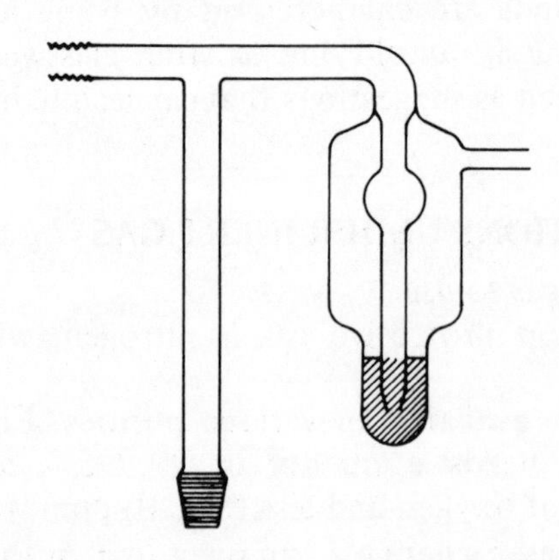

A pressure-equalising dropping funnel is particularly useful as the inert gas will then also protect the contents of the funnel.

*How to de-gas solvents*

The simplest procedure is to boil the solvent for some time while passing through the inert gas and then to distill in an inert atmosphere.

Small quantities (5–10 ml) of solvent can also be freed from interfering gases as follows: Place the solvent in a flask fitted with a ground glass tap, cool to −70°C, and evacuate the flask using a high vacuum pump. Warm to room temperature. Repeat the process twice.

*How to transfer small quantities (<20 ml) of oxygen or water sensitive liquids*

For this purpose it is particularly convenient to use a syringe with a needle (whose length may reach that of a normal NMR tube). The syringe and needle are dried at 125°C in a drying cabinet and allowed to cool under protective gas, or in a desiccator. Stock bottles and reaction vessels are fitted with serum caps and solutions are transferred by syringe. Ideally, the material removed should be replaced by inert gas admitted using a syringe needle and bubble valve. If it is not possible to equalise pressure with inert gas, one can compensate by filling the syringe – before piercing the serum cap – with a quantity of inert gas equal in volume to the liquid to be withdrawn. For larger quantities one could use a dry pipette fitted with a detachable 'propipette'.

For the transfer of larger quantities of oxygen or water sensitive solutions or liquids, long flexible stainless steel syringe needles are used:

The needle is introduced via a septum into the gas space in the vessel above the sensitive liquid and is flushed using a slight excess pressure of protective gas. Then the free end of the needle is inserted via another septum into the receiving flask, which is equipped for the passage of protective gas. The end of the needle in the vessel containing the sensitive liquid is then pushed below the surface of the fluid: the slight excess pressure causes the liquid to be tranferred.

Reagent bottles with screw tops may conveniently be opened under an inverted funnel through whose stem protective gas is being passed Fig. (a) p. 102, or in a plastic bag flushed with gas. With higher boiling solvents, such as THF, a U-shaped glass tube with protective gas passing through it can be hung over the neck of the bottle.

*How to transfer solid substances which are moisture sensitive*

The easiest way to transfer solids is to follow the procedure described above for reagent bottles with screw tops. To add a moisture sensitive solid to a reaction mixture in portions a flask, constructed as in Fig. (b) p. 102, with a ground glass joint, is ideal. It is important that the shape is chosen to ensure that in position (a) no material can trickle into the reaction mixture, while in position (b) all the remaining material can be tapped lightly into the reaction flask, making sure that even in position (b) the flask doesn't get in the way of a condenser or stirrer fitted to the reaction vessel.

In place of such a flask one can successfully use an ordinary flask connected to the reaction flask by a flexible tube of large internal diameter.

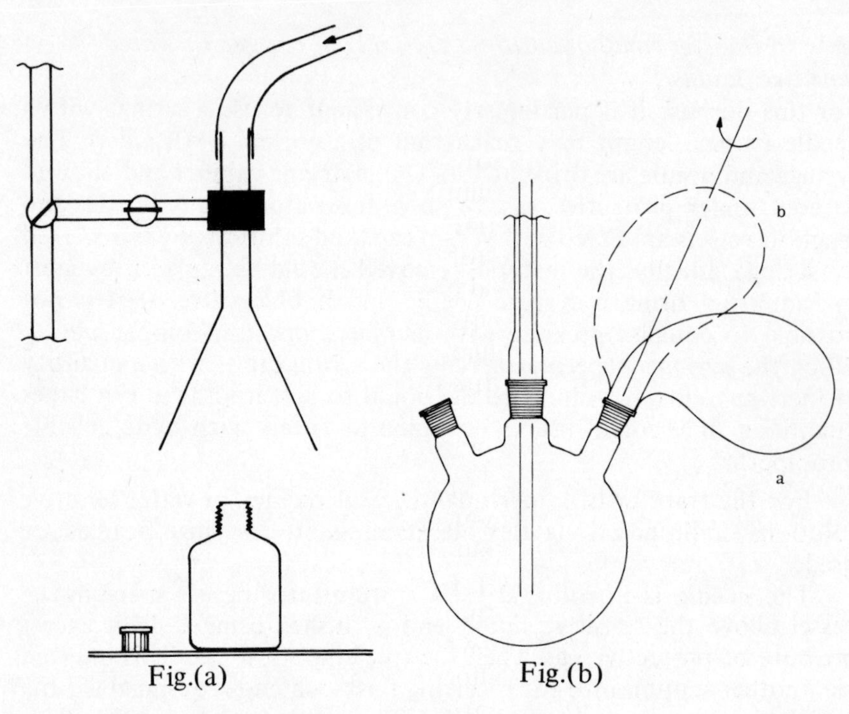


Fig.(a) Fig.(b)

*How to filter under protective gas*

The 'pressure filter' shown in the figure has proved extremely suitable for this purpose (Ref. [1]). In order to filter, one introduces inert gas through (a) and then inserts a loose-fitting plug at (b). If one is mainly interested in the filtrate, it is advisable to use a two-necked flask

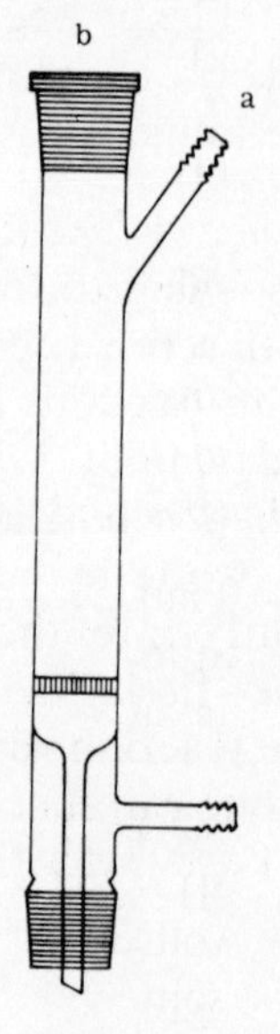

fitted with a ground glass tap – it is flushed with inert gas before the sinter is removed – or to fit between the sinter and the flask a straight adaptor through which protective gas can be passed when the sinter is removed. Sensitive suspensions can be transferred directly into the 'pressure filter' from a flask by using a tap fitted with ground glass cones at each end. For filtration of a reaction mixture under protective gas, see also Ref. [2].

In less critical cases filtration can be carried out under protective gas in a conventional 'glove bag'.

*How to crystallise air or moisture sensitive substances.*
For this purpose the filter with the side tube at the top, as shown in the figure, is suitable (Ref. [3]). A collecting flask can be attached to the apparatus. The sensitive material is introduced as a suspension and, as in the case of the 'pressure filter' described above, is separated from the solvent by forcing protective gas at moderate pressure through (a). To crystallise the material, gas is then passed in through the collecting flask at (b), serving, in addition, to stir the solution when the solvent is added. If required the whole filter can be immersed in a cold-bath.

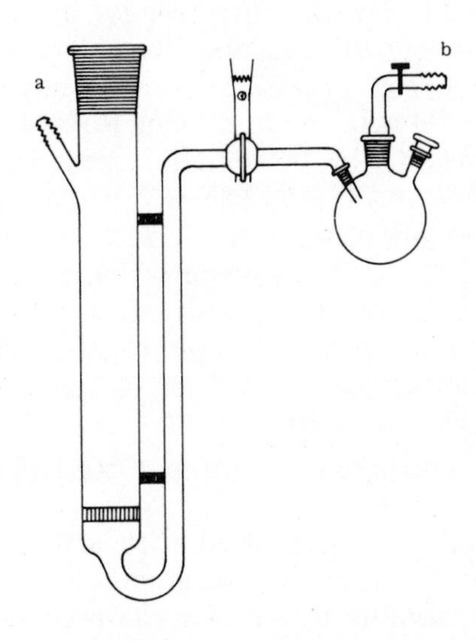


In the simplest cases, crystallisation under protective gas can be carried out in a two-necked flask fitted with a ground glass tap which serves to introduce the gas. In this case the mother liquor will need to be removed by pipetting or siphoning.

*How to store moisture or air sensitive substances*
For this purpose one may use a two-necked flask fitted with a serum cap and a ground glass tap. Ampules fitted with serum caps from which fluids and solutions can be removed when needed using a syringe are better for longer term storage. However, for the best long-term results, ampules sealed under inert gas or after evacuation are often essential.

To seal ampules successfully, the ampule neck needs to have been constricted beforehand. Cool the contents adequately, evacuate, close the vacuum stop-cocks, and seal using a fine flame.

*Manipulation of some commonly used reagents*
*Lithium, alkyls, $BH_3$ in THF, diethylaluminium cyanide in toluene, diisobutyl-aluminium hydride in hexane:* For transfer from the usual commercial bottles (with screw tops) see above.

*Lithium aluminium hydride:* The powdered reagent is extremely sensitive to moisture. It is advisable to open the can in which it is supplied under an inverted funnel through which protective gas is being passed and to weigh it out only in closed vessels. Before closing the stock bottle flush it with protective gas. Once opened, the cans should have their plastic lids firmly replaced and should then be stored in a desiccator filled with protective gas. It may be more convenient to use the commercially available 1g tablets: in crystalline form the reagent is not pyrophoric even in moist air. $LiAlH_4.2THF$ may be obtained as a 3.5M solution in toluene Ref. [4].

*Potassium hydride:* The powdered reagent is usually supplied commercially under paraffin oil. The sedimented powder is pipetted out, under protective gas, using a pipette with a wide orifice. The adhereing paraffin oil can be removed by filtration and washing with pentane, using a 'pressure filter' (see above). Dry, oil-free potassium hydride flows easily, but in the presence of moisture is spontaneously flammable. It is advantageous to use a pre-weighed sinter so that one can determine the weight of dry reagent.

*Lithium sand:* Powdered lithium in a dry and flowing condition can be prepared as follows:

In a cylindrical vessel fitted with a vibratory stirrer, an argon inlet, and a drying tube (see diagram), lithium wire (1 cm pieces) and the 2% of sodium needed for the preparation of lithium alkyls are heated in boiling dry tetralin (b.p. 207°C). The molten lithium is stirred vigorously and then the heating and vibrator are switched off and the suspension is cooled. The tetralin is carefully covered with a layer of dry pentane; the light lithium rises to the top of the pentane layer. Using a pipette most of the brownish tetralin solution beneath

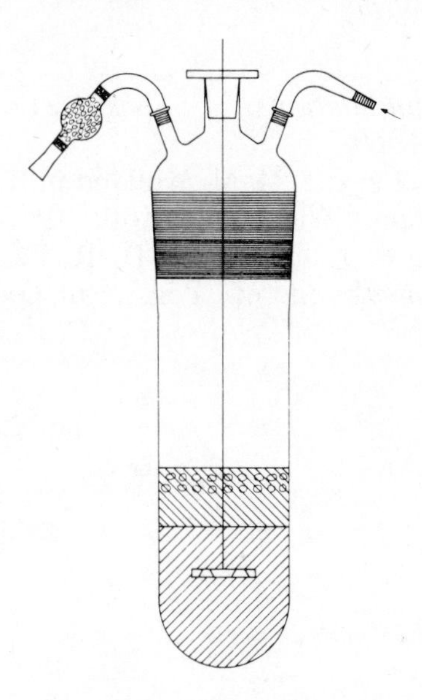

the pentane is removed and the lithium sand is washed several times with more pentane. To obtain dry lithium, the pentane is removed by filtration under pressure of protective gas in a 'pressure filter' (see above). Protective gas is passed through the lithium sand to dry it.

*Sodium sand:* see, e.g. *Org. Synth., Coll. Vol.* V, p. 1090.

*Potassium:* see, e.g. *Org. Synth., Coll. Vol.* IV, p. 134. Potassium cannot be freed from adhering oil by the usual drying paper, as it may spontaneously catch fire in the process. Potassium cut under liquid paraffin or xylene should be dipped briefly into dry hexane and immediately transferred to a reaction vessel flushed with protective gas.

## REFERENCES

[1] Houben-Weyl, *Methoden der Organischen Chemie,* 4th ed., Vol 1/2, p. 361.

[2] *Organic Syntheses,* **59**, 124.

[3] This filter was designed by Dr. J. Schreiber, ETH Zurich. See also K. Hafner, A. Stephen, and C. Bernhard, *Liebigs Ann. Chem.*, **650**, 54 (1961).

[4] (a) J. S. Pizey, *Synthetic Reagents,* Ellis Horwood, Chichester, 1974.

(b) G. Brendel, *Chem. and Eng. News,* 11 May, 1981, p.3.

## BIBLIOGRAPHY

D. F. Shriver, *The Manipulation of Air-Sensitive Compounds,* McGraw-Hill, New York, 1969.

G. W. Kramer, A. B. Levy, & M. M. Midland in H. C. Brown, *Organic Synthesis via Boranes,* Wiley, New York, 1975.

D. D. Perrin, W. L. F. Armarego, & D. R. Perrin, *Purification of Laboratory Chemicals,* 2nd ed., Pergamon, Oxford, 1980.

CHAPTER 12

# Hints on the Synthesis of Isotopically Labelled Compounds

## 1) SYNTHESIS WITH STABLE ISOTOPES

Compounds containing isotopes in other than the proportions in which they occur naturally may be required, for example, as n.m.r. solvents or for the elucidation of reaction mechanisms.

Isotopically labelled target molecules may in most cases be prepared from simple starting materials which may be obtainable with different degrees of isotopic enrichment:–

| $^{2}H$ | $^{12}C$ | $^{13}C$ | $^{15}N$ | $^{17}O$ | $^{18}O$ |
|---|---|---|---|---|---|
| $D_2O$ | $BaCO_3$ | $BaCO_3$ | $NaNO_3$ | $H_2O$ | $CO_2$ |
| $D_2$ | $CH_3OH$ | $CH_3OH$ | $NaNO_2$ | $O_2$ | $BaCO_3$ |
| $LiBD_4$ | $CH_3I$ | $CH_3I$ | $NH_4Cl$ | $CO_2$ | $H_2O$ |
| $B_2D_6$ | NaCN | KCN | $N_2H_4$ | CO | $Me_2SO$ |
| $CD_3OD$ | | $^{*}CH_3CO_2Na$ | CuCN | | |
| $C_6D_6$ | | $CH_3{}^{*}CO_2Na$ | $(NH_2)_2CO$ | | |
| $CD_3I$ | | C (amorph.) | | | |
| $(CD_3)_2CO$ | | $CH_2O$ | | | |

In addition, an increasing number of compounds, particularly those of bio-organic interest, are becoming available with isotopic labels.

For the manipulation of labelled reagents in synthesis, the following tips may be used to supplement the material in *Hints on the Synthesis of Organic Compounds.*

– Experience shows that if a labelled compound is commercially available, a good practical synthesis is often *not* found in the literature.

– Many labelled compounds, especially those containing $^{2}H$, can lose much of their isotopic enrichment by exchange reactions. Syntheses must be devised to avoid exchange. Thus, methanol-$d_1$ may be prepared by the hydrolysis of dimethyl carbonate with $D_2O$.

– The limiting factor in most syntheses is the cost of the labelled starting material. Design syntheses to maximise incorporation of the isotopic label, preferably by insertion of the labelled fragment as late as possible in the synthetic route.
– Compounds containing $^{2}H$, $^{13}C$, $^{15}N$, and $^{18}O$ can be handled in ordinary glass apparatus.

The preparation of isotopically labelled compounds requires special measures to be taken in order to satisfy the above considerations.
– First check all preparations using *unlabelled* material, on the same scale, at the same place, and using the same apparatus.
– In some cases (and *always* in preparations using radioactive labels) it is advantageous to use a 'sandwich technique':– Start the reaction with *ca.* 10 mole % of unlabelled reactant. After a suitable time (dictated by the rate at which the reaction proceeds), add the labelled sample (1–50 mole %) and allow the reaction to proceed for the usual period. *Then* add sufficient unlabelled material to complete the transformation.

*Always* test the 'sandwich' reaction in *advance* using only unlabelled materials, but following *exactly* the conditions to be employed. The time is well spent!
– If $^{2}H$ is to be incorporated at a high level, it is usually worth using specific reactions. Exchange reactions with $D_2O$ at equilibrium produce significantly lower deuterium levels than were present in the original $D_2O$.

**Special Hints**

*Deuteriated compounds:* In many compounds deuterium may be readily exchanged by protons. In such cases it is essential to protect the materials from any contact with moisture. (See *Some Tips on Handling Water or Air Sensitive Substances).*

*Deuterium oxide:* This is very hygroscopic: a single transfer from one vessel to another can reduce the deuterium content by 0.015%. It is best to store it under a small positive pressure of argon.

*Deuterium gas:* Used for deuterium incorporation via the reduction of double and triple bonds. Note the following points:
– Use proton-free catalysts: wash with $D_2O$ and dry in an atmosphere of $D_2$ gas.
– Use no protic solvents for the reduction for fear that the catalyst may also promote deuterium-proton exchange!

– The gas reserve and pressure equilibration vessels should be filled with liquid paraffin, not water.

*Sodium borodeuteride* ($NaBD_4$): Protons are exchanged for deuterium on contact with water; atmospheric moisture must be rigorously excluded. It is an advantage to add *ca.* 0.5% deuterium oxide to sodium borodeuteride for reductions in aprotic solvents if high deuterium incorporation is required: heat the borodeuteride to 120°C and then allow to cool in a dry desiccator in which a small vial containing the requisite amount of $D_2O$ has been placed. The optimal reaction conditions, which lead to the utilisation of all four deuterium atoms, should be determined empirically using unlabelled material.

*Lithium borodeuteride* ($LiBD_4$): Intrinsically more reactive and more sensitive to moisture than $NaBD_4$ and should therefore always be transferred and reacted under protective gas. For many purposes it can be prepared *in situ* by the reaction of sodium borodeuteride and lithium chloride in dry diglyme.

*Lithium aluminium deuteride* ($LiAlD_4$): Very sensitive to moisture. If loss of isotope is unacceptable, it is essential to work in a dry box.

*Deuterodiborane* ($B_2D_6$): Best prepared *in situ* from dry sodium borodeuteride and boron trifluoride-etherate.

*Deuterotrifluoroacetic acid* ($CF_3COOD$): Mainly used to catalyse proton-deuteron exchange, especially in n.m.r spectroscopy. It is very poisonous and hygroscopic and should only be handled in serum ampules from which it can be withdrawn using a dry syringe.

$^{13}C$ *compounds:* The majority of $^{13}C$-labelled starting materials (see list above) are one-carbon units. Unlike deuterium, $^{13}C$ exchange is often negligible in practice. However, one needs to avoid decarboxylations, decarbonylations, and other cleavage processes, which may reduce $^{13}C$ incorporation.

## 2) SYNTHESIS WITH RADIOISOTOPES ($^{14}C$, $^{13}H$)

BEFORE EMBARKING ON REACTIONS WITH RADIOACTIVE SUBSTANCES WITH ACTIVITIES ABOVE THE LEGAL LIMITS, MAKE SURE YOU KNOW THE RELEVANT RULES AND REGULATIONS AND THAT YOU HAVE BEEN PROPERLY AUTHORISED. RADIOACTIVE COMPOUNDS SHOULD ONLY BE USED IN SPECIALLY EQUIPPED RADIOCHEMICAL LABORATORIES.

WHENEVER SUCH COMPOUNDS ARE BEING MANIPULATED, WEAR GLOVES, LABORATORY COATS, AND SAFETY SPECTACLES. ONE OFTEN NEEDS QUITE DIFFERENT TECHNIQUES FROM THOSE USED IN CONVENTIONAL CHEMISTRY. SUBSTANCES ARE FREQUENTLY TRANSFERRED USING VACUUM LINES. ALL REACTION MUST BE CARRIED OUT AT ALL TIMES ON SAFETY TRAYS, SO THAT THERE IS NO DANGER OF THE LABORATORY BEING CONTAMINATED BY RADIOACTIVITY. THERE ARE STRICT REGULATIONS GOVERNING THE DISDISPOSAL OF RADIOACTIVE WASTES.

For further details, consult the works listed in the *Bibliography* at the end of this section.

The following notes on syntheses with radioisotopes, should be read in conjunction with *Synthesis with Stable Isotopes.*

– Except for a few special, highly-enriched tritium compounds, e.g. $^3H_2$ gas, even so-called 'high-level' radioactively labelled substances contain only a small proportion of labelled molecules diluted with large quantities of compound containing only the stable isotopes. This is especially important in the case of tritium (very high isotope effect: $k_H/k_T$ = 7:1 – 14:1) – any step in which complete reaction with the labelled atom does not occur will result in a major loss of label; e.g. transfer of $^3H$ (cation) to a carbanion using $[^3H]H_2O$ results in preferential reaction with $^1H$ – in such cases use $[^3H]CF_3CO_2H$ (no enolisable H), prepared by the reaction of equimolar amounts of $[^3H]H_2O$ and $(CF_3CO)_2O$, to ensure complete transfer of $^3H$. On the other hand, $^3H$ transfer by equilibration (e.g. tritiation of an aromatic ring by prolonged reaction under acidic conditions) may use $[^3H]H_2O$ efficiently.

– Always use a 'sandwich technique' for the incorporation of label from a radioactive precursor.

– Always test the 'sandwich' reaction first using unlabelled material.

– The radioactive product should be purified after each step to constant specific activity; in the case of double labelling, purify to a constant ratio of activity (e.g. $^{14}C$:$^3H$).

– For $^{14}C$ (half-life 5730 years) calibration with respect to a fixed date is unnecessary, but it is important for $^3H$ (half-life 12.35 years), and vital in many other cases (e.g. $^{32}P$, half-life 14.3 days).

### 3) DETERMINATION OF SPECIFIC ACTIVITY IN SAMPLES SINGLY OR DOUBLY LABELLED WITH $^{14}C$ AND $^3H$

The measurement of radioactivity using scintillators is based on the following principles:

Decomposition of a radionuclide results in the excitation of a number of solvent molecules which in turn transfer their excitation energy to scintillator molecules. The light emission from these molecules is measured using a photomultiplier. The number of impulses measured depends on the intensity of the photo-emission, which is in turn proportional to the radioactivity present. The $\beta$-emission from $^{3}H$ at 18 KeV is at a lower evergy than the 156 KeV radiation for $^{14}C$. It is significant for the measurement of the actual activity that the energy spectrum $^{14}C$ overlaps the region for $^{3}H$. Accordingly, measurements for $^{3}H$ must be corrected if $^{14}C$ is determined simultaneously. One must also take into account partial quenching of the emission by chemical processes and by colour (whether intrinsic or as impurities).

For the determination of specific activity one needs:
– A small glass or plastic tray for carrying radioactive samples. This tray should be kept exclusively for one's personal use.
– Dark screw-cap bottles for each of the scintillator solutions, 'aqueous' (usually contains dioxan) and 'organic' (usually in toluene). Commercial formulations are available, but for special applications one may make up one's own 'cocktail'. These solutions should be measured out only by pouring, not by pipette. Accuracy should be *ca.* 1%.
– $^{14}C$ and $^{3}H$ standard solutions: it is convenient to use *n*-hexadecane of standard activity (one each for $^{14}C$ and $^{3}H$). The reagent bottles should be dated (especially for $^{3}H$, which loses *ca.* 5% of its activity in a year; a correction needs to be calculated).
– Two labelled 1 ml glass syringes for weighing out standards. Use separate syringes for $^{14}C$ and $^{3}H$ and never mix them up! The syringes should be stored in labelled test tubes. It is undesirable to wash them out after use.
– For weighing use disposable Al boats – use once only.
– Disposable Al discs are used to avoid contamination of balances during weighing.
– Counting vials with screw caps are used. Polyethylene ones are destroyed after use; glass vials may contain residual radioactivity even after washing.
– 'Primary' solvents, in which samples are dissolved for counting, include: water methanol, ethanol, benzene, dioxan. Chlorinated hydrocarbons and ketones are unsuitable. If necessary, these solvents are measured out in graduated 1 ml pipettes.

**Counting Radioactive Materials ($^{14}C$ and $^{3}H$)**
In the following section it is assumed that we are dealing essentially with single determinations, in which internal standardisation is

desirable. For routine measurements, external standardisation is adequate provided quenching is uniform.

(A) *Low-level samples* ($< 10\ \mu Ci/mmol$)

1. Weigh sample on a 6-figure balance in a weighed disposable Al boat. Usually *ca.* 1 mg is used, but aim for an amount of compound which will give 10 000–20 000 counts per min; as a minimum one needs a counting rate 10 × background. Use an Al disc to protect the balance pan from contamination.
2. Transfer the weighed sample to a (polyethylene) counting vial.
3. Dissolve the sample in 'primary' solvent (if water $<0.2$ ml, if organic solvent $<0.5$ ml). Solution must be complete.
4. Add scintillator solution ('aqueous' if water is primary solvent) to a standard total volume, in the range $7.0 \pm 0.2$ to $10.0 \pm 0.2$ ml. Swirl carefully (do not shake). Check that there is no precipitation. Close cap.
5. Count sample, preferably $> 10\ 000$ counts. Repeat twice. (If counts fall, the sample may be precipitating; if they rise, solution may not have been complete.)
6. Add standard(s).
   (a) *Single isotope.*
      (i) Weigh out by difference on a 5 figure balance an appropriate amount of standard *n*-hexadecane in a 1 ml syringe kept for this purpose. The amount (20-80 mg) should be chosen to double the counting rate. Add standard to the counting vial making sure the needle does not touch the side of the vial.
      (ii) Count the resulting solution twice.
   (b) *Double labelling with* $^{14}C$ *and* $^{3}H$
      In this case the order of addition of standards is crucial.
      (i) Weigh out and add $^{14}C$ standard.
      (ii) Count resulting solution twice.
      (iii) Weigh out and add $^{3}H$ standard.
      (iv) Count the solution containing both standards twice.

(B) *High-Level Samples* ($>10\mu Ci/mmol$)

HIGH-LEVEL SAMPLES SHOULD BE HANDLED ONLY IN A PROPER RADIOCHEMICAL LABORATORY.

Only diluted solutions should be taken to the counter.

1. Weight out *ca.* 1 mg of the sample, using a 6-figure balance, into a volumetric flask. Take the greatest care to avoid contamination of the balance.
2. Make up to the mark using 'primary' solvent.

3. Transfer a measured volume (pipette or syringe) to a counting vial and proceed as in A.4–6 above.

If the approximate activity is known, choose the volume transferred to give a counting rate of 10 000–20 000 per min. For samples of very high levels of radioactivity, double dilution may be needed.

## Calculation of Counting Efficiency, Decomposition Rate and Specific Activity

The calculation is shown for a sample doubly labelled with $^{14}C$ and $^{3}H$. The numbers refer to the 'counting sheet' displayed overleaf and the points A.1–6 above.

II.1.1 The *total counting rate of the sample* without any added standard is determined in both the $^{3}H$ and the $^{14}C$ channels [see A.5].

1.2 The increased counting rate after addition of the $^{14}C$ standard is measured in both channels [A.6(b) (i) and (ii)].

1.3 By subtracting 2–1 one obtains the $^{14}C$ counting rate for the added standard.

1.4 The counting efficiency in the $^{14}C$ channel is given by:

$$\frac{\text{counting rate of } ^{14}\text{C standard (II.1.3)}}{\text{decomposition rate of } ^{14}\text{C standard (I.2.5)}}$$

The counting efficiency for $^{14}C$ in the $^{3}H$ channel ( * ) is given by:

$$\frac{\text{counting rate of } ^{14}\text{C standard in } ^{3}\text{H channel}}{\text{decomposition rate of } ^{14}\text{C standard (I.2.5)}}$$

It should have a value $<0.1$.

II.2.1–4 After addition of the $^{3}H$ standard, the counting rates and efficiencies [A.6(iii) and (iv)] are calculated analogously to II.1. The $^{3}H$ efficiency in the $^{14}C$ channel (**) should be $<0.001$.

III.1.2. The background should be counted immediately after the sample, especially if the activity is weak.

III.1.4 The $^{14}C$ decomposition rate† is calculated from the $^{14}C$ counting rate of the sample after correcting for background:

†The small correction due to $^{3}H$ decomposition being counted in the $^{14}C$ channel may be disregarded.

(*), (**) See Counting Sheet p 114

$$^{14}\text{C decomposition rate} = \frac{\text{corrected } ^{14}\text{C counting rate (III.1.3)}}{^{14}\text{C efficiency in } ^{14}\text{C channel (II.1.4)}}$$

III.2.1 The $^{3}$H counting rate is corrected for $^{14}$C decompositions from the $^{14}$C decompostion rate of the sample and the counting efficiency for $^{14}$C in the $^{3}$H channel (III.1.4 X II.1.4[*]).

III.2.3 $^{3}$H decomposition rate is given by:

$$\frac{\text{effective } ^{3}\text{H counting rate}}{^{3}\text{H counting efficiency in } ^{3}\text{H channel}}$$

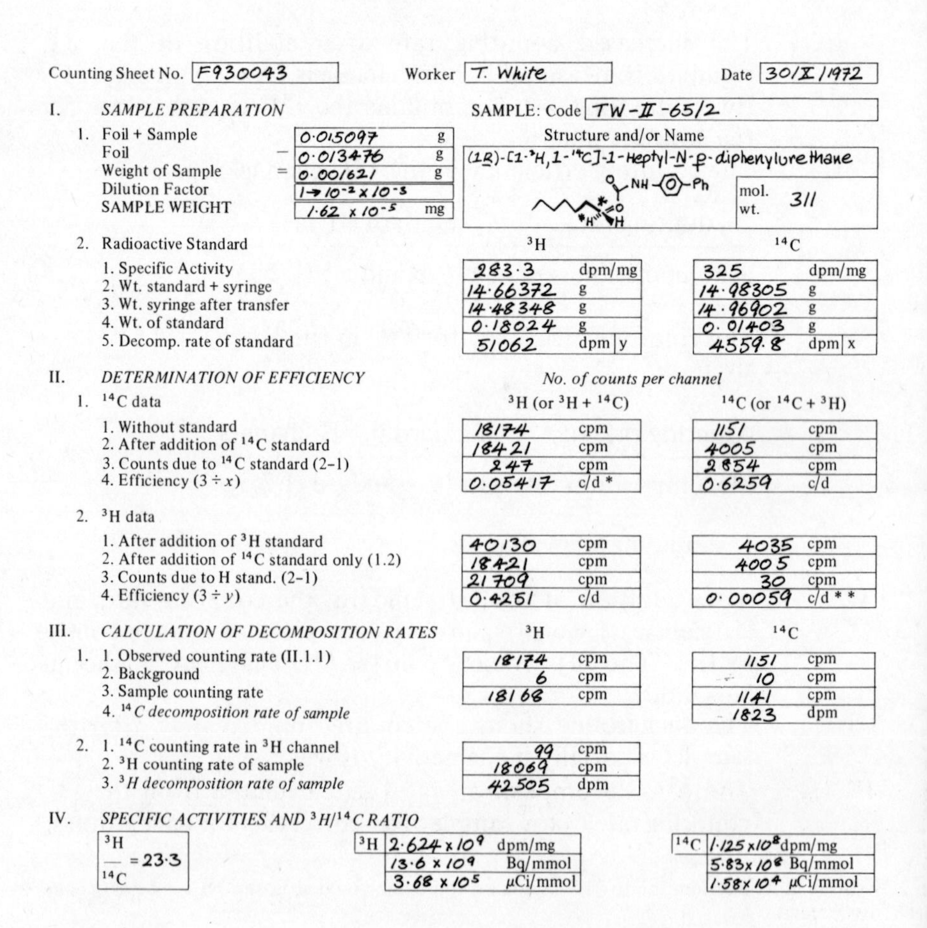

Counting Sheet No. F930043 Worker T. White Date 30/X/1972

I. *SAMPLE PREPARATION* SAMPLE: Code TW-II-65/2

| 1. | | | Structure and/or Name |
|---|---|---|---|
| Foil + Sample | | 0·015097 g | (1R)-[1-$^{3}$H,1-$^{14}$C]-1-Heptyl-N-p-diphenylurethane |
| Foil | − | 0·013476 g | |
| Weight of Sample | = | 0·001621 g | |
| Dilution Factor | | 1→10⁻² x 10⁻³ | |
| SAMPLE WEIGHT | | 1·62 x 10⁻⁵ mg | mol. wt. 311 |

2. Radioactive Standard

| | $^{3}$H | $^{14}$C |
|---|---|---|
| 1. Specific Activity | 283·3 dpm/mg | 325 dpm/mg |
| 2. Wt. standard + syringe | 14·66372 g | 14·98305 g |
| 3. Wt. syringe after transfer | 14·48348 g | 14·96902 g |
| 4. Wt. of standard | 0·18024 g | 0·01403 g |
| 5. Decomp. rate of standard | 51062 dpm y | 4559·8 dpm x |

II. *DETERMINATION OF EFFICIENCY* *No. of counts per channel*

| 1. $^{14}$C data | $^{3}$H (or $^{3}$H + $^{14}$C) | $^{14}$C (or $^{14}$C + $^{3}$H) |
|---|---|---|
| 1. Without standard | 18174 cpm | 1151 cpm |
| 2. After addition of $^{14}$C standard | 18421 cpm | 4005 cpm |
| 3. Counts due to $^{14}$C standard (2–1) | 247 cpm | 2854 cpm |
| 4. Efficiency (3 ÷ $x$) | 0·05417 c/d * | 0·6259 c/d |

| 2. $^{3}$H data | | |
|---|---|---|
| 1. After addition of $^{3}$H standard | 40130 cpm | 4035 cpm |
| 2. After addition of $^{14}$C standard only (1.2) | 18421 cpm | 4005 cpm |
| 3. Counts due to H stand. (2–1) | 21709 cpm | 30 cpm |
| 4. Efficiency (3 ÷ $y$) | 0·4251 c/d | 0·00059 c/d ** |

III. *CALCULATION OF DECOMPOSITION RATES*

| 1. | $^{3}$H | $^{14}$C |
|---|---|---|
| 1. Observed counting rate (II.1.1) | 18174 cpm | 1151 cpm |
| 2. Background | 6 cpm | 10 cpm |
| 3. Sample counting rate | 18168 cpm | 1141 cpm |
| 4. *$^{14}$C decomposition rate of sample* | | 1823 dpm |

| 2. | $^{3}$H |
|---|---|
| 1. $^{14}$C counting rate in $^{3}$H channel | 99 cpm |
| 2. $^{3}$H counting rate of sample | 18069 cpm |
| 3. *$^{3}$H decomposition rate of sample* | 42505 dpm |

IV. *SPECIFIC ACTIVITIES AND $^{3}$H/$^{14}$C RATIO*

$^{3}$H/$^{14}$C = 23·3

| $^{3}$H | $^{14}$C |
|---|---|
| 2·624 x 10⁹ dpm/mg | 1·125 x 10⁸ dpm/mg |
| 13·6 x 10⁹ Bq/mmol | 5·83 x 10⁸ Bq/mmol |
| 3·68 x 10⁵ μCi/mmol | 1·58 x 10⁴ μCi/mmol |

*Notes*

– If the efficiency is too low because of colour quenching, the sample can be 'bleached': for example, benzoyl peroxide is added and the closed sample vial irradiated with UV for several hours. As long as the vessel is sealed, no activity is lost even if the chemical nature of the radioactive molecules is completely altered.

– In all cases where chemical modification in the counting vial is required, use glass vials, not polyethylene. This applies also to $^{40}K$.

– If the decomposition rate for $^{14}C$ in a doubly labelled sample is much greater than that for $^{3}H$, measure the $^{3}H$ rate first.

– Some units:

| | | |
|---|---|---|
| Decomposition rate: | disintegrations per minute | dpm |
| Counting rate: | counts per minute | cpm |
| Yield: | counts per disintegration | c/d |
| Specific activity: | | dpm/mg |

– Some definitions:

1 Curie (1 Ci): the quantity of a substance which gives $3.7 \times 10^{10}$ dps (disintegrations per second)

1 Becquerel (1Bq): 1 dps

## BIBLIOGRAPHY

A. F. Thomas, *Deuterium Labelling in Organic Chemistry,* Meredith Corporation, 1971.

D. G. Ott, *Synthesis with Stable Isotopes of Carbon, Nitrogen and Oxygen,* Wiley, 1981.

A. Murray & D. L. Williams, *Organic Synthesis with Isotopes,* Wiley Interscience, 1958.

*Journal of Labelled Compounds and Radiopharmaceuticals*

*A Guide to Users of Labelled Compounds,* 2nd edition, 1979, and various *Radiochemical Reviews* (especially numbers 11, 14, and 17) issued free by the Radiochemical Centre, Amersham, Bucks. HP7 9LL, England. Useful free literature is also available from other suppliers.

I. Kirshenbaum, *Physical Properties and Analysis of Heavy Water,* McGraw-Hill, 1951.

D. L. Horrocks, *Applications of Liquid Scintillation Counting,* Academic Press, 1974.

CHAPTER 13

# Disposal and the Destruction of Dangerous Materials

Waste disposal is a problem that urgently requires a comprehensive solution. Because waste from research laboratories often contains substances whose danger can easily be underestimated, particular care is necessary in dealing with it. (The following recommendations are those in force at the ETH, Zurich and conform to local legal requirements: the procedures are valid, but may not be suited to all localities).

## 1. CLASSIFICATION OF WASTE MATERIAL

*Dangerous materials should be clearly labelled as such.*

*A. Waste*

Harmless, solid organic or inorganic residues which can be safely incinerated or dumped. Must contain no category C or D materials.

*B. Organic Solvents and Chemicals.*

Must contain no category C or D materials.

(a) Chlorinated solvents ($CH_2Cl_2$, $CHCl_3$, $CCl_4$, $C_6H_5Cl$, etc.).
(b) Other solvents (including dissolved organic materials as long as they do not react with water or with each other; the total chlorine content should be less than 5%).
(c) Organic chemicals, especially waste oils and greases.

*C. Dangerous materials.*

Poisonous, corrosive, oxidising, malodorous, or pyrophoric substances or solutions, or ones that are explosive or inflammable, e.g. alkali metals, hydrides, organometallic compounds, poisonous gases, cyanides, acid halides, diazo compounds, nitro compounds, *N*-nitrosamines, peroxides and peroxy acids, poisonous phosphorus compounds, chlorosulphonic acid, oleum, mercury, heavy metal salts, etc. Must contain no category D substances.

*D. Radioactive substances.*

## 2. WASTE DISPOSAL

*A. Waste.*

Put in sealed bags into suitable containers, collect, and remove to dump. In principle, waste should contain no 'chemicals'.

*B. Organic Solvents and Chemicals*

One should use the least chlorinated solvent possible (e.g. $CH_2Cl_2$ rather than $CHCl_3$ rather than $CCl_4$). Chlorinated solvents are expensive and disposal is not easy. Accordingly, they should be recovered, distilled, and re-used whenever possible, e.g. solvents stripped by rotary evaporator after recrystallisation or chromatography may often be used again. Waste chlorinated solvents should *never* be stored in contact with waste acetone.

'Washing' acetone can be distilled for re-use, as long as the main contaminants are removed.

Many used solvents as well as organic liquids, oils and fats can be disposed of by incineration. Specialised firms deal with such problems.

*C. Dangerous Materials*

Dangerous materials should be rendered harmless without delay, preferably in the laboratory. Take heed of any special precautions published. Neutralise acids and bases (especially if more than 1 mol).

- Take the utmost care. Don't rush or panic.
- Never work alone.
- Always wear safety glasses, and if necessary rubber gloves.
- Use an empty fume hood.
- Use the correct procedure and add reagents slowly.
- In general, excess reagents should be used for decomposition.
- Stir, have a cooling bath ready, work in an inert atmosphere (nitrogen, argon, etc.).
- Make sure the decomposition reaction is complete.
- Under no circumstances keep residues for more than two months.
- Never pour dangerous materials down the drain.
- Dangerous materials that cannot be neutralised on the spot should be carefully sealed in suitable containers, labelled, and sent for disposal to specialised firms.

*D. Radioactive Substances*

As under C, but greater care is needed. Sealing of containers for specialised disposal is essential.

## SPECIAL INSTRUCTIONS FOR THE DESTRUCTION OF DANGEROUS SUBSTANCES

|  |  |
| --- | --- |
| Alkali metal hydrides or amides, sodium 'sand'. | Suspend in dry dioxan (tetrahydrofuran, ether), add ethanol or propan-2-ol slowly with stirring and swirling until hydrogen evolution ceases. Then add water slowly until the solution is clear. Wash away. |
| Alkali metal borohydrides | Dissolve in methanol, dilute well with water, add acid and allow to stand. Some boranes are exceptionally poisonous, so be sure to work in a fume cupboard. Neutralise and wash away. |
| Acid chlorides or anhydrides phosphorus oxychloride phosphorus pentachloride, thionyl chloride, sulphuryl chloride. | Add a large excess of water or 10% NaOH with stirring, then neutralise. (For $P_2O_5$ see p.125). Wash away. |
| Catalysts (Ni, Cu, Fe, noble metals, filters or celite pads containing tese catalysts – they are often pyrophoric when dry). | Never place in waste bins. Never suck completely dry when filtering. Small amounts (< 1 g) can be washed down the drain using plenty of water. Larger amounts should be sent in sealed, labelled containers for disposal. |
| Chlorine, bromine, sulphur dioxide! | Absorb in NaOH solution. Neutralise and wash away. |
| Chlorosulphonic acid, concentrated sulphuric or hydrochloric acid, 'oleum'. | Add dropwise to stirred ice or ice-water then neutralise and wash away. |
| Dimethyl sulphate | Add dilute NaOH or $NH_4OH$ carefully with stirring, then neutralise and wash away. |
| Hydrogen sulphide, mercaptans, thiophenols, cyanogen chloride or bromide, hydrogen cyanide, phosphines, solutions of sulphides or cyanides. | Oxidise with hypochlorite (NaOCl, bleach) – 1 mol mercaptan requires *ca.* 2 ℓ bleach (17% Cl, 9 mol 'active Cl') – 1 mol cyanide requires *ca.* 0.4 ℓ bleach. Use nitrite paper to establish that bleach is present in excess (pH > 7). Wash away. |
| Heavy metals and their salts. | Precipitate as insoluble compounds (carbonates, hydroxides, sulphides etc.), then specialised firms. |
| Lithium aluminium hydride | Suspend in dry dioxan (tetrahydrofuran, ether), add ethyl acetate dropwise (cool if necessary), then add water until hydrogen evolution ceases. Wash away (for large amounts neutralise first) |

| | |
|---|---|
| Mercury | Pick up metal quickly, then to specialised firms (or recover). Commercial adsorbents (such as MERCUROSORB) can be used to clear up spillages safely and rapidly.<br>Amalgams – specialised firms.<br>Salts and solutions – precipitate as sulphide, filter and then to specialised firms. |
| Organolithium compounds | Dissolve in dioxan (tetrahydrofuran, ether), add ethanol slowly until hydrogen evolution ceases, then add water followed by dilute acid until a clear solution is obtained. Wash away. |
| Peroxide or peroxyacids in solution | Reduce (Fe(II) salts, bisulphite) in aqueous, acidic solution, then netralise and wash away. |
| Phosgene (or phosgene solutions in non-chlorinated organic solvents) | Absorb in 15% NaOH solution. Wash away. |
| Potassium (spontaneously inflames in air, burns explosively on contact with water, highly reactive – always cut under petrol/hexane) | Drop in small pieces into dry t-butanol, add SVR ethanol (*not* methanol) carefully, stir or leave until entirely dissolved, then neutralise. Wash away. |
| Sodium | Drop in small pieces into ethanol or propan-2-ol, leave until dissolved, add water slowly to give a clear solution, then neutralise and wash away. |
| Sulphur trioxide | Pass into conc. $H_2SO_4$, then destroy as for conc. acids. |

## BIBLIOGRAPHY

*Guide for Safety in the Chemical Laboratory,* 2nd edition, Manufacturing Chemists Association, Van Nostrand, Reinhold, N.Y., 1972.

*Hazards in the Chemical Laboratory,* L. Bretherick ed., 3rd edition, Royal Society of Chemistry, London, 1981.

## CHAPTER 14

# Purification and Drying of Solvents

Purity, as applied to a solvent, is not an absolute concept. It is impossible to prove the total absence of a particular impurity: a negative result for any chemical or physical test merely shows that the concentration of a substance is below the detection limit of the test. In practice, an organic chemist is less interested in this philosophical 'absolute' purity than in the suitability of a solvent for a particular application or reaction.

Careful purification (and drying – water is also an impurity) of solvents usually proves worthwhile. In particular, it is advisable to use freshly distilled solvents at all times.

In this section are described the main drying agents, and procedures are given for purification and drying of the most common solvents, together with their properties. Quite apart from anything discussed here, the demands on the purity of solvents for particular applications should always be considered anew. These procedures can only act as guides in this matter.

Improper handling of most solvents can lead to serious health hazards. Accordingly, the threshold limit values (TLV) for each solvent, i.e. the maximum concentrations that one can encounter with safety in a laboratory or factory have been given.

### TLV – THRESHOLD LIMIT VALUES

The TLV is the maximum concentration in the atmosphere of a material in the form of gas, vapour, or dust which, as far as is at present known, can be tolerated under working conditions by the overwhelming majority of healthy people, without any damage to health, for up to 8–9 hours a day and 45 hours a week, even over long periods.

A TLV forms a basis for deciding whether or not a particular concentration of a dangerous material constitutes a hazard under working conditions. In addition to the toxicity of the inhaled substance, other factors, e.g. corrosive action, sensitisation, irritation, absorption through the skin, are taken into account in assessing a TLV.

A TLV is not a definite boundary between safe and dangerous situations. It gives no information concerning the hazards associated with shorter exposures at higher concentrations.

For chemical carcinogens no 'safe concentration' can be stated (at least on current knowledge). As a result, no TLV is quoted for most carcinogens.

**Some selected TLV values** (in ppm)

| | | | |
|---|---|---|---|
| Ammonia | 50 | Formic acid | 5 |
| Aniline | 5 | Hydrogen chloride | 5 |
| *iso*Amyl alcohol | 100 | Hydrogen cyanide | 10 |
| *p*-Benzoquinone | 0.1 | Hydrogen sulphide | 10 |
| Boron trifluoride | 1 | Iodine | 0.1 |
| Bromine | 0.1 | Lead | 0.1 |
| Carbon dioxide | 5000 | Maleic anhydride | 0.2 |
| Carbon monoxide | 50 | Osmium tetroxide | 0.0002 |
| Chlorine | 0.5 | Ozone | 0.1 |
| Chlorinated biphenyls (possibly carcinogenic) | 0.05 | Phenol | 5 |
| Cyclohexene | 300 | Phosgene | 0.1 |
| *N, N*-Diethylamine | 25 | Sulphur dioxide | 2 |
| Diborane | 0.1 | Tetraethyllead | 0.01 |

## A SOME DRYING AGENTS

### Alumina

$Al_2O_3$

M. W. 101.96 Capacity maximum 10%

(a) Removal of peroxides from ethers and hydrocarbons: 100 g alumina (activity I, basic) in a 15 mm diameter column will remove peroxides by adsorptive filtration from 1000 cm$^3$ diethyl ether, 400 cm$^3$ di-isopropyl ether, 1000 cm$^3$ tetralin, 100 cm$^3$ dioxan. Basic alumina activity Super I is approximately twice as efficient.

The peroxides are adsorbed but not decomposed. Used alumina should therefore be washed away with water.

(b) Purification of hydrocarbons for U.V. spectroscopy: see e.g. 'Hexane' P. 137.

(c) Removal of water from organic solvents.

Economical only for solvents which have previously been dried and distilled. In that case 150 g alumina (activity I, basic) should suffice to remove residual moisture from 100–1100 $cm^3$ of solvents containing less than 0.01% water. It is particularly suitable for hydrocarbons and chlorinated hydrocarbons, as well as for many ethers and esters. In the process, peroxides and other polar impurities are removed.

(d) Removal of ethanol used to stabilise chloroform: see under 'Chloroform' p. 130.

**Barium Oxide**
$BaO$
M. W. 153.34
T.L.V. (dust) 0.5 mg $m^{-3}$

A relatively vigorous drying agent used for ethyl acetate and recommended for organic bases[1]. To be freshly crushed; powdery reagent is probably no longer active.

**Calcium chloride**
$CaCl_2$ Residual water: 0.25 mg per ℓ dry air.
M. W. 110.99 Capacity: *ca.* 90%.

Cheap, slow, and not particularly efficient. Takes up water to form the hexahydrate below 30°C. Suitable for pre-drying hydrocarbons, alkyl halides, ethers, and many esters.

It melts on taking up excess water. Drying tubes should be used only once, and tested before use for effectiveness!

**Calcium oxide**
$CaO$ Residual water: 0.2 mg per ℓ dry air.
M. W. 56.08

Cheap. Suitable for drying low-boiling alcohols and amines. Heat together under reflux for an hour or two, then distil the solvent.

**Calcium sulphate** – semihydrate (Drierite, Sikkon)
$CaSO_4.0.5\,H_2O$ Residual water: 0.004 mg per ℓ dry air
M. W. 145.15 Capacity: *ca.* 7%

Powerful drying agent, almost completely utilisable. Suitable for almost all organic liquids and gases.

**Lithium aluminium hydride (LAH)**
$LiAlH_4$
M. W. 37.95

Very efficient at drying hydrocarbons and ethers. These solvents must be pre-dried as LAH reacts vigorously with water.

Dangerous: it decomposes above 150°C; even using an inert atmosphere a distillation over LAH should never be allowed to go to dryness! Decompose the excess reagent (see procedure p. 118).

**Magnesium**
Mg
Atomic weight 24.31

Used for making 'superdry' alcohols: see under 'Methanol' p. 137.

**Magnesium sulphate**
$MgSO_4$ — Residual water: 1 mg per ℓ dry air
M. W. 120.37 — Capacity: *ca.* 100% (formation of hepta-hydrate).

Suitable for drying almost all compounds, including acids and their derivatives, aldehydes, and ketones. Standard medium for drying solutions.

For drying exceptionally acid-sensitive compounds use sodium sulphate or a basic drying agent.

**Molecular sieve**

Molecular sieve is a synthetic crystalline aluminium silicate. After the water of crystallisation has been removed (i.e. in the active condition) it has a large number of cavities associated with pores of a closely-defined molecular diameter (3,4,5, or 10 Å, according to type). These cavities can then be re-occupied, but only by molecules having critical van der Waals radii less than the pore diameters. For example:

| | | | | | | | |
|---|---|---|---|---|---|---|---|
| $H_2$ | 2.4 Å | $CO_2$ | 2.8 Å | $NH_3$ | 3.8 Å | $C_2H_4$ | 4.25 Å |
| $H_2O$ | 2.6 Å | $N_2$ | 3.0 Å | $Cl_2$ | 8.2 Å | $C_2H_6$ | 4.44 Å |
| $O_2$ | 2.8 Å | CO | 3.2 Å | $C_2H_2$ | 2.4 Å | n-alkanes | 4.89 Å |

The maximum capacity of molecular sieve is about 20%. Its bulk density lies between 50–70 g per 100 $cm^3$.

Molecular sieve has the advantage of being capable of regeneration at will with effectively no loss of capacity. It should always be reactivated before use by heating in an oven at 300-350°C either *in vacuo* or in a stream of nitrogen or argon. In the absence of a suitable drying oven the following procedure will do: mix used molecular sieve with plenty of water, filter, and dry well. Dry for a few hours in a drying cupboard at 150°C, then under high vacuum at 200°C (silicon oil bath) overnight. When dealing with large quantities it is advantageous to interrupt the vacuum drying after two hours and flush with argon to ensure that the granules dry adequately. This treatment gives molecular sieve with a residual water content less than 0.5%.

Drying organic solvents is most efficient if the pre-dried, distilled, peroxide-free solvent is percolated through a column of molecular sieve (dynamic drying). A 250 g molecular sieve column (25 mm diameter, 600 mm high) will dry 10 litres of the following solvents to a purity better than 0.002% residual water at 3 litres an hour: diethyl ether, diisoprophyl ether, tetrahydrofuran, dioxan, benzene, toluene, cyclohexane, dichloromethane, chloroform, carbon tetrachloride, ethyl acetate (all these with 4 Å sieve), and acetonitrile (3 Å sieve).

Molecular sieve quickly takes up atmospheric moisture and so should be handled rapidly. In order to obtain the solvent free of dust it is advantageous to use molecular sieve activated by the procedure given above just before use. Dried solvents can be stored over molecular sieve (*ca.* 10 g $\ell^{-1}$).

**Potassium carbonate**
$K_2CO_3$
M. W. 138.21

Suitable for preliminary drying of organic bases at room temperature.

**Potassium hydroxide**
KOH Residual water: 0.002 mg per ℓ dry air
M. W. 56.11

Suitable for drying organic bases. Melts as it takes up water (use in a drying tower). It is much more efficient than sodium hydroxide.

Dangerous: work carefully, wear safety glasses!

**Silica gel** (Kieselgel) Capacity: *ca.* 35%
Use in desiccators and drying tubes. Best to use the granules containing an indicator that is blue when dry and pink when saturated with water.

**Phosphorus pentoxide**
$P_2O_5$ Residual water: less than 0.000025 mg per ℓ dry air
M. W. 141.94

Very rapid and efficient. One of the best drying agents, except that the surface becomes syrupy as water is taken up and this hinders further uptake. This problem is overcome in various patent preparations (e.g. Fluka phosphorus pentoxide drying agent – 75% $P_2O_5$ in an inert carrier, Merck Sicapent) which remain particulate even after 100% uptake of water.

Phosphorus pentoxide is suitable for drying saturated and aromatic hydrocarbons, anhydrides, nitriles, alkyl and aryl halides, and carbon disulphide. It is *not* suitable for alcohols, amines, acids, aldehydes, or ketones.

Care is required in the decomposition of large amounts of excess $P_2O_5$. A practical method is to add it in small portions to a large quantity of ice, and subsequently to neutralise with base.

**Sodium**
Na
Atomic weight 22.99

Suitable for drying ethers, tertiary amines, and saturated or aromatic hydrocarbons. Only pre-dried solvents should be dried with sodium.

Sodium has been largely superseded as a drying agent by metal hydrides, which are easier to use, equally efficient, and simpler to decompose.

Excess sodium must be destroyed (see p. 119).

**Sodium hydride**
NaH
M. W. 23.998

Used usually in the form of a dispersion in oil for drying ethers and hydrocarbons. Sodium hydride may ignite spontaneously on contact with water and burns explosively. Excess hydride must be destroyed (see p. 118).

**Sodium sulphate**
$Na_2SO_4$ Residual water: 12 mg per ℓ dry air
M. W. 162.04 Capacity: *ca.* 75%

Useful for preliminary drying at room temperature of sensitive compounds, e.g. acids, aldehydes, ketones, halides. The theoretical capacity (formation of decahydrate) is unattainable. Somewhat less efficient than magnesium sulphate.

**Sulphuric acid**
$H_2SO_4$
M. W. 98.08

Suitable for inert neutral or acidic gases.

Available in 'granulated form' (25% inert carrier), with or without an indicator (e.g. Merck Sicacide).

## B. SOLVENTS

(For destruction of sodium, sulphuric acid, phosphorus pentoxide, etc. used in drying and purification see pp. 118–9).

### Acetone

| | | |
|---|---|---|
| $CH_3COCH_3$ | m.p. −94.7°C | $d_4^{20}$ 0.78998 |
| M. W. 58.081 | b.p. 56.29°C | dielectric const. (25°) 20.70 |
| T.L.V. 1000 ppm | odour threshold 200–450 ppm | flash point (flame induced) −30°! |

Relatively harmless. Recommended for cleaning glassware. Can be recovered by distillation in favourable cases (discard grossly impure fractions into solvent residues can). Completely miscible with methanol, ethanol, ether, water, etc. No azeotrope with water.

On contact with basic or acid reagents acetone forms condensation products.

*Purification:* distillation over $CaSO_4.0.5H_2O$ (e.g. Sikkon), *ca.* 100 g $\ell^{-1}$, Molecular sieve – 3Å (see p 123).

### Acetic acid (glacial acetic acid).

| | | |
|---|---|---|
| $CH_3CO_2H$ | m.p. 16.66°C | $d_4^{25}$ 1.04366 |
| M. W. 60.053 | b.p. 117.90°C | dielectric const. (20°C) 6.15 |
| T.L.V. 10 ppm | No water azeotrope | |

Completely miscible with water. Extremely hygroscopic.

*Purification:* For many purposes it is sufficient to purify by crystallisation. Cool in an ice-bath until it solidifies and decant the mother liquor when about 80% of the acid has crystallised; repeat twice. Can be dried over phosphorus pentoxide (but some acetic anhydride may be formed).

### Acetic anhydride

| | | |
|---|---|---|
| $(CH_3CO)_2O$ | m.p. −73.10°C | $d_4^{15}$ 1.08712<br>$d_4^{30}$ 1.06911 |
| M. W. 102.091 | b.p. 140.00°C | dielectric const. (19°C) 20.7 |

T.L.V. 5 ppm! (Poisons lungs, irritates mucous membranes)

Hydrolyses very slowly in water at 0°C and pH7.

*Purification:* Stand for a few hours over phosphorus pentoxide (100 g $\ell^{-1}$), decant, stand over the same amount of potassium carbonate for a few hours, filter, and distil in the absence of moisture at 100 torr. It can also be purified by careful fractional distillation using an efficient column.

**Acetonitrile**

| | | |
|---|---|---|
| $CH_3CN$ | m.p. −43.835°C | $d_4^{25}$ 0.7766 |
| M. W. 41.053 | b.p. 81.60°C | dielectric const. (20°C) 37.5 |
| T.L.V. 40 ppm. | Water azeotrope: 76.7° | (84.2% acetonitrile) |

Completely miscible with ethanol, ether, water; slightly soluble in hydrocarbons.

*Purification:* distillation over 1% (by weight) of phosphorus pentoxide, then from *ca.* 5% (by weight) anhydrous potassium carbonate.

For u.v. spectroscopy: fractionally distil commercial acetonitrile using an efficient column (e.g. 150 cm vacuum-jacketed packed 'Normag' column). Treat the constant boiling fraction under reflux for 1 hour with 20 g potassium permanganate and 20 g anhydrous potassium carbonate per litre, then distil rapidly. Add the distillate (which is basic and smells of ammonia) dropwise to conc. sulphuric acid until the acid endpoint. Decant from any inorganic precipitate and distil in an inert atmosphere using an efficient column (see above) and discarding the forerun (5–10%).

**Aniline** (aminobenzene)

| | | |
|---|---|---|
| $C_6H_5NH_2$ | m.p. −5.98°C | $d_4^{25}$ 1.01750 |
| M. W. 93.129 | b.p. 184.40°C | dielectric const. (20°C) 6.89 |

T.L.V. 5 ppm – also by skin adsorption, so avoid skin contact entirely.

*Purification:* Fractional distillation under water-pump vacuum, if necessary after treatment with barium oxide.

**Anisole** (phenyl methyl ether)

| | | |
|---|---|---|
| $C_6H_5OCH_3$ | m.p. −37.5°C | $d_4^{20}$ 0.996 |
| M. W. 108.15 | b.p. 153.75°C | dielectric const. 4.33 |

*Purification:* Fractional distillation under water-pump vacuum, or dry over calcium chloride and then distil from sodium.

**Benzene**

| | | |
|---|---|---|
| $C_6H_6$ | m.p. 5.533°C | $d_4^{25}$ 0.87370 |
| M. W. 78.115 | b.p. 80.100°C | dielectric const. 2.275 |

T.L.V. 8 ppm: long term carcinogen (leukaemia). Handle only in fume-hood. When ever possible use (less dangerous) toluene or xylene instead. Water azeotrope: 69.25°C (91.17% benzene).

At 20° it is saturated by 0.06% water.
*Purification:* (a) Distillation, discarding 10% forerun. (b) Distillation from sodium hydride dispersion (0.5 g $\ell^{-1}$). (c) Shake repeatedly with portions (5% by volume) of conc. sulphuric acid, wash well with water, dry over calcium sulphate (100 g $\ell^{-1}$), decant, distil from sodium hydride (10.5 g $\ell^{-1}$). If this last method has to be used for toluene or xylene, prolonged contact with sulphuric acid should be avoided as sulphonation of these latter hydrocarbons is facile.

**Butan-1-ol**

| | | |
|---|---|---|
| $CH_3CH_2CH_2CH_2OH$ | m.p. −88.62°C | $d_4^{25}$ 0.8060 |
| M. W. 74.124 | b.p. 117.66°C | dielectric const. (25°C) 17.51 |

T.L.V. 100 ppm. Water azeotrope: 92.70°C (57.5% butanol)
At 25°C, 7.45% soluble in water, and dissolves 20.5% water itself.
*Purification:* Reflux over calcium oxide, or treat with sodium/dibutyl phthalate (*cf.* 'ethanol').

**Butan-2-ol**

| | | |
|---|---|---|
| $CH_3CH_2CH(OH)CH_3$ | m.p. −114.7°C | $d_4^{25}$ 0.8026 |
| M. W. 74.124 | b.p. 99.95°C | dielectric const. (25°C) 16.56 |

T.L.V. 100 ppm. Water azeotrope: 87.0°C (73.2% butanol).
At 20°, 12.5% soluble in water; at 25°C it dissolves 44.1% water.
*Purification:* Distillation.

tert-**Butanol** (2-methylpropan-2-ol)

| | | |
|---|---|---|
| $(CH_3)_3COH$ | m.p. 25.82°C | $d_4^{30}$ 0.7757 |
| M. W. 74.124 | b.p. 82.42°C | dielectric const. (25°C) 12.47 |

T.L.V. 100 ppm. Water azeotrope 79.9°C (88.24% butanol).
Ternary azeotrope with water/benzene 67°C.
Completely miscible with water, ethanol, ether.
As it is solid at room temperature, it will solidify in a condenser: use a circulating pump and water from a thermostatic bath at *ca.* 30°.
*Purification:* (a) Reflux over calcium oxide (*cf.* 'ethanol'). (b) Crystallise (by cooling) in the absence of moisture. (c) Pre-dry, then distil from sodium (1%).

iso-**Butanol** (2-methylpropan-1-ol)

| | | |
|---|---|---|
| $(CH_3)_2CHCH_2OH$ | m.p. −108°C | $d_4^{25}$ 0.7978 |
| M. W. 74.124 | b.p. 107.66°C | dielectric const. (25°C) 17.93 |

T. L. V. not known Water azeotrope 89.8°C (67% butanol)

At 25°C, 10% soluble in water, and dissolves 16.9% water itself. *Purification:* (a) Reflux over calcium oxide (*cf.* 'ethanol'), then distil using an efficient column. (b) Pre-dry, then treat with magnesium or calcium (*cf.* 'methanol').

**Butan-2-one** (methyl ethyl ketone, MEK)
$CH_3CH_2COCH_3$ m.p. −86.69°C $d_4^{25}$ 0.7997
M.W. 72.108 b.p. 79.64°C dielectric const. (20°C) 18.51
T.L.V. 200 ppm. Water azeotrope: 73.41°C (88.73% ketone).
At 20°C, 24% soluble in water, and dissolves 10% water itself.
*Purification:* see 'acetone'.

**t-Butyl methyl ether**
$(CH_3)_3COCH_3$ m.p. −108.65°C $d_4^{25}$ 0.7352
M.W. 88.15 b.p. 55.2°C
T.L.V. Not known Water azeotrope 52.6° C (4% water).
Ternary azeotrope with methanol/water. Stable in neutral or alkaline solution, but decomposed by mineral acids. Minimal autoxidation in storage.
*Purification:* see 'diethyl ether'.

**Carbon disulphide**
$CS_2$ m.p. −111.57°C $d_4^{15}$ 1.27005 $d_4^{30}$ 1.24817
M.W. 76.139 b.p. 46.225°C dielectric const. (20°C) 2.641
T.L.V. 20 ppm. It is extremely flammable. Mixtures of carbon disulphide and air have flash points (flame induced) of about −30°, ignition temperature (at a hot surface) is 100° (steam bath!). Water azeotrope: 42.6°C (97.2% $CS_2$).
At 20°C, it is saturated by less than 0.005% water.
*Purification:* Distillation over a little phosphorus pentoxide ($< 10$ g $\ell^{-1}$).

**Carbon tetrachloride**
$CCl_4$ m.p. −22.95°C $d_4^{25}$ 1.58439
M. W. 153.823 b.p. 76.75°C dielectric const. (20°C) 2.238
T.L.V. 10 ppm! Chronic exposure causes liver damage. Do not inhale vapour. Handle in a fume hood. Avoid whenever possible. Water azeotrope: 66°C (95.9% carbon tetrachloride). At 24°C, it is saturated by 0.010% water. Carbon tetrachloride is not flammable.

*Purification:* Reflux over phosphorus pentoxide (5 g ℓ$^{-1}$) for 30 minutes, then distil. For spectroscopic use, this distillate should be filtered through alumina (basic activity I) (see 'chloroform').

**Chlorobenzene**

$C_6H_5Cl$ m.p. −45.58°C $d_4^{20}$ 1.10630
M. W. 112.560 b.p. 131.687°C dielectric const. (25°C) 5.621

T.L.V. 75 ppm. Water azeotrope 90.2°C (71.6% chlobenzene).
At 25°C it is saturated by 0.0327% water.

*Purification:* Shake repeatedly with portions (*ca.* 5 cm$^3$ ℓ$^{-1}$) of conc. $H_2SO_4$. Wash until neutral, dry ('Sikkon', 50 g ℓ$^{-1}$), decant, distil over $P_2O_5$ (5 g ℓ$^{-1}$). If required free of all traces of acid, pass through an alumina (basic, activity I) column (100 g ℓ$^{-1}$) (*cf.* 'chloroform'). Drying agent: molecular sieve 4 Å.

**Chloroform**

$CHCl_3$ m.p. −63.55°C $d_4^{25}$ 1.47988
M. W. 119.378 b.p. 61.152°C dielectric const. (20°C) 4.806

T.L.V. 10 ppm. It is anaesthetic and can cause liver damage. Exceptionally damaging to eyes (Caution using in syringes e.g. for i.r. spectroscopy). Water azeotrope: 56.12°C (97.8% chloroform).

At 23°C it is saturated by 0.072% water.

Not flammable. In light it reacts with oxygen to form phosgene ($COCl_2$), chlorine, hydrogen chloride, etc., and commercial chloroform is therefore stabilised by the addition of *ca.* 1% ethanol.

*Ethanol-free chloroform* (e.g. for i.r. spectroscopy):
adsorptive filtration through alumina (basic, activity I); a 50 g column will give *ca.* 70 cm$^3$ chloroform containing less than 0.005% ethanol and free of traces of water or acid. The first 25 cm$^3$ should be collected and put onto the column again as the forerun tends to contain excess water; the heat of adsorption liberated when the column begins operation reduces its efficiency and it only reaches its full drying power when it has cooled. Stabiliser-free chloroform can be kept in the refrigerator in a dark bottle, but not longer than two weeks.

*Purification:* Distillation over phosphorus pentoxide (*ca.* 5 g ℓ$^{-1}$).
Danger: chloroform may react explosively with strong bases or alkali metals!

**Cyclohexane**

$C_6H_{12}$ m.p. 6.544°C $d_4^{25}$ 0.77389
M. W. 84.162 b.p. 80.725°C dielectric const. (20°C) 2.023

T.L.V. 300 ppm Water azeotrope: 68.95°C (91% cyclohexane).
At 20°C, it is saturated by 0.01% water.
*Purification:* see 'hexane'.

**Decalin** (decahydronaphthalene, bicyclo [4,4,0] decane) – mixture of isomers.

$C_{10}H_{18}$ m.p. −124 ± 2°C $d_4^{25}$ 0.8789
M. W. 138.255 b.p. 191.70°C dielectric const. (25°C) 2.1542

T.L.V. less than 100 ppm. May cause eczema.
*Purification:* (a) Heat under reflux with sodium wire (*ca.* 1%) for a few hours in an inert atmosphere (nitrogen, argon), then distil. (b) Filter through alumina (basic, activity I; *ca.* 100 g $\ell^{-1}$). Forms peroxides with oxygen in light, so store under nitrogen or argon. Can be dried over molecular sieve 4Å but peroxides are not then removed (*cf.* 'diethyl ether').

**1,2-Dichloroethane** (ethylene chloride)

$ClCH_2CH_2Cl$ m.p. −35.66°C $d_4^{24}$ 1.2453
M. W. 98.960 b.p. 83.483°C dielectric const. (25°C) 10.36

T.L.V. 20 ppm., possibly carcinogenic. Water azeotrope 72°C (91.8% dichloroethane). Azeotropes with methanol, ethanol, propan-2-ol, carbon tetrachloride.

At 20°C it is saturated by 0.15% water.

*Purification:* Distil over phosphorus pentoxide (5 g $\ell^{-1}$), then filter through alumina (basic, activity I, 50 g $\ell^{-1}$). If all traces of olefin need to be removed, before distilling shake several times with portions of concentrated sulphuric acid (5% by volume), wash until neutral, and dry over magnesium sulphate.

**Dichloromethane** (methylene chloride)

$CH_2Cl_2$ m.p. −95.14°C $d_4^{25}$ 1.31678
M. W. 84.933 b.p. 39.75°C dielectric const. (25°C) 8.93

T.L.V. 200 ppm. (much less poisonous than chloroform, T.L.V. 10 ppm., or carbon tetrachloride, T.L.V. 10 ppm.)

Water azeotrope 38.1°C (98.5% dichloromethane).

At 25°C saturated by 1.30% water.

*Purification:* Reflux for 30 minutes over phosphorus pentoxide (5 g $\ell^{-1}$), then distil. Traces of acid in the distilled solvent can be removed by filtration through alumina (basic, activity, I, 50 g $\ell^{-1}$). Danger: *never* allow dichloromethane to come into contact with alkali metals or strong bases – violent reaction!

**Diethyl ether** (ether)

| | | |
|---|---|---|
| $CH_3CH_2OCH_2CH_3$ | m.p. −116°C | $d_4^{25}$ 0.70760 |
| M. W. 74.124 | b.p. 34.55°C | dielectric const. (20°C) 4.335 |

T.L.V. 400 ppm. Ether vapour is highly flammable with a flash point (flame induced) of −45°C and an ignition temperature (at a hot surface) of 180°C; it is heavier than air and it will creep along a bench top for a considerable distance.

At 25°C, 6.04% soluble in water, and dissolves 1.468% water itself.

Commercial ether contains variable amounts of water, ethanol, and peroxides.

*Purification:* Add sodium hydride (*ca.* 1 g $\ell^{-1}$) slowly to the ether, reflux for 30 minutes, then distil. Ether suitable for almost all applications can be prepared by a second distillation, this time in an inert atmosphere (nitrogen, argon) from lithium aluminium hydride (0.5 g $\ell^{-1}$).

In the presence of light, ether combines with oxygen to form peroxides which are concentrated on distillation and can explode violently. Accordingly, unpurified ether should *never* be taken to dryness. The presence of peroxides can be detected as follows: shake 3 $cm^3$ ether with a solution of potassium iodide (200 mg) in hydrochloric acid (I N, 10 $cm^3$) – a brown colour (liberated iodine) indicates that peroxides are present.

Ether should be stored in the absence of light. Bottles should always be used within a few days of being opened (avoid large air volume above the ether).

Ether can also be purified by distillation from sodium and filtration through alumina (basic, activity I) (100 g $\ell^{-1}$). The spent alumina should not be heated, but washed away with plenty of water, as the peroxides are adsorbed, but not destroyed. Alternatively, dry over molecular sieve 4 Å (which does not remove peroxides – a mixture of molecular sieve and alumina may be used if necessary).

**Diethyl carbonate**

| | | |
|---|---|---|
| $(CH_3CH_2O)_2CO$ | m.p. −43.0°C | $d_4^{25}$ 0.96926 |
| M. W. 118.134 | b.p. 126.8°C | dielectric const. (20°C) 2.820 |

T.L.V. not known

**Diethylene glycol dimethyl ether** (diglyme)
$CH_3OCH_2CH_2OCH_2CH_2OCH_3$

| | | |
|---|---|---|
| | m.p. −64°C | $d_4^{25}$ 0.9440 |
| M. W. 134.177 | b.p. 159°C (decomposition) | dielectric const. (20°C) 7 |

T.L.V. not known. It is poisonous.

Hygroscopic. Totally miscible with water and most of the common solvents.

*Purification:* (a) Fractional distillation under reduced pressure (water-pump, b.p. 62–63°C/15 torr). (b) Treat with several batches of fresh sodium wire until no further reaction is detected, decant, fractionate in inert atmosphere. (c) Stand for several hours over calcium hydride, then reduced pressure fractional distillation.

**Dimethoxyethane** (monoglyme, ethylene glycol dimethyl ether)

| | | |
|---|---|---|
| $CH_3OCH_2CH_2OCH_3$ | m.p. −58°C | $d_4^{20}$ 0.8665 |
| M. W. 90.12 | b.p. 93.0°C | dielectric const. (25°C) 7.20 |

Health hazard. Water azeotrope (89.9% monoglyme)
Completely miscible with water and many organic solvents.
*Purification:* Treat with several batches of fresh sodium wire until no further reaction is detected, decant, fractionate in inert atmosphere (nitrogen, argon).

**N,N-Dimethylformamide** (DMF, formdimethylamide)

| | | |
|---|---|---|
| $(CH_3)_2NCHO$ | m.p. −60.43°C | $d_4^{25}$ 0.94397 |
| M. W. 73.095 | b.p. 153.0°C | dielectric const. (25°C) 36.71 |

T.L.V. 20 ppm. No water azeotrope

At 150°C it slowly decomposes to give dimethylamine and carbon monoxide. Decomposition is base-catalysed and will proceed at room temperature e.g. in the presence of sodium hydroxide. DMF is completely miscible with water and many organic solvents. It is a good solvent for salts.

*Purification:* Mix DMF (250 g), benzene (30 g) [or cyclohexane (30 g)], and water (12 g). Distil at 140°C to remove impurities together with the water and benzene. Cool (exclude moisture) and then distil under reduced pressure (water-pump). Store in the dark.

**Dimethylsulphoxide** (DMSO)

| | | |
|---|---|---|
| $(CH_3)_2SO$ | m.p. 18.54°C | $d_4^{25}$ 1.0958 |
| M. W. 78.134 | b.p. 189°C | dielectric const. (25°C) 46.68 |

T.L.V. not known but definitely not harmless. Readily adsorbed through the skin, and it carries many substances through with it. Avoid *all* contact with the skin. No water azeotrope.
*Purification:* Dry over calcium sulphate or molecular sieve 4Å, then distil under reduced pressure (water-pump, b.p. 75°C/12 torr).

**1,4–Dioxan** (dioxan, *p*-dioxan)
$C_4H_8O_2$ m.p. 11.80°C $d_4^{25}$ 1.02797
M. W. 88.107 b.p. 101.32°C dielectric const. (25°C) 2.209
T.L.V. 50 ppm.; possibly carcinogenic
Water azeotrope: 87.82°C (82% dioxan)

Completely miscible with water.

Forms peroxides (for detection, see 'diethyl ether').

*Purification:* Pre-dry over solid potassium hydroxide, decant, distil over sodium. Store in inert atmosphere (nitrogen, argon). Can be dried over molecular sieve 4Å. Molecular sieve will not remove peroxides.

**Di-isopropyl ether**
$(CH_3)_2CHOCH(CH_3)_2$ m.p. −85.5°C $d_4^{25}$ 0.7182
M. W. 102.187 b.p. 68.3°C dielectric const. (25°C) 3.88
T.L.V. 500 ppm Water azeotrope 62.2°C (95.5% ether)

At 20°C it is saturated by 0.57% water.

*Purification:* see 'diethyl ether'
Danger: Forms peroxides with oxygen very readily!

**Ethanol** (ethyl alcohol)
$CH_3CH_2OH$ m.p. −114.1°C $d_4^{25}$ 0.78504
M. W. 46.070 b.p. 78.29°C dielectric const. (25°C) 24.55
T.L.V. 1000 ppm. Water azeotrope 78.14°C (96% ethanol). Ternary azeotrope with water/cyclohexane 62.1°C (17% ethanol, 76% cyclohexane, 7% water). This ternary azeotrope does not separate into two phases on cooling. It can be used to dry ethanol or for esterification.

Completely miscible with benzene, chloroform, ether, acetone, water.

*Purification:* Treat commercial absolute alcohol (which may contain 0.01% water or more) with sodium (*ca.* 7 g $\ell^{-1}$) in the absence of moisture. When ethoxide formation is complete, add *ca.* 30 g diethyl phthalate, reflux for 2 h, and distil.

A less efficient purification: Reflux SVR (96%) ethanol over *ca.* 25% calcium oxide for 15 h and then distil.
Further drying: Magnesium (see under 'methanol'), calcium sulphate.

**Ethylene chloride** – see '1,2-Dichloroethane'

**Ethylene glycol** (ethane-1,2-diol)

| | | |
|---|---|---|
| $HOCH_2CH_2OH$ | m.p. −13°C (glass) | $d_4^{25}$ 1.1100 |
| M. W. 62.029 | b.p. 197.3°C | dielectric const. (25°C) 37.7 |

T.L.V. – negligible (involatile). However, it is poisonous (*cf.* methanol), unlike glycerine. No water azeotrope.
*Purification:* Reduced pressure distillation. Dry over sodium sulphate. Inclined to bump in boiling.

**Ethylene glycol dimethyl ether** – see '1,2-Dimethoxyethane'

**Ethylene glycol monoethyl ether** (Ethyl cellosolve, 2-ethoxyethanol)

| | | |
|---|---|---|
| $CH_3CH_2OCH_2CH_2OH$ | m.p. below −90°C | $d_4^{25}$ 0.92520 |
| M. W. 90.123 | b.p. 135.6°C | dielectric const. (24°C) 29.6 |

T.L.V. 200 ppm.
*Purification:* Distillation

**Ethylene glycol monomethyl ether** (methyl cellosolve, 2-methoxyethanol)

| | | |
|---|---|---|
| $CH_3OCH_2CH_2OH$ | m.p. −85.1°C | $d_4^{25}$ 0.96024 |
| M. W. 76.096 | b.p. 124.6°C | dielectric const. (25°C) 16.23 |

T.L.V. 25 ppm.
*Purification:* Distillation.

**Ethyl acetate**

| | | |
|---|---|---|
| $CH_3CO_2CH_2CH_3$ | m.p. −83.97°C | $d_4^{25}$ 0.89455 |
| M. W. 88.107 | b.p. 77.114°C | dielectric const. (25°C) 6.02 |

T.L.V. 400 ppm. Water azeotrope: 70.38°C (91.53% ester)
At 25°C 8.08% soluble in water, and it takes up 2.94% water itself.
*Purification:* (a) Reflux for 2 hours over barium oxide (5 g $\ell^{-1}$), then distil. (b) Dry over phosphorus pentoxide or molecular sieve 4 Å, then distil.

**Ethyl formate**

| | | |
|---|---|---|
| $HCO_2CH_2CH_3$ | m.p. −79.4°C | $d_4^{25}$ 0.924 |
| M. W. 74.080 | b.p. 54.15°C | dielectric const. (25°C) 7.16 |

T.L.V. 100 ppm  Water azeotrope: 52.6°C (95% ester).

At 20° it dissolves 17% water.

*Purification:* Dry over anhydrous magnesium sulphate, decant, then distil over phosphorus pentoxide (*ca.* 10 g $\ell^{-1}$).

Further drying agents: anhydrous sodium sulphate or potassium carbonate (*not* calcium chloride, with which it forms an addition compound).

**Formamide**

| | | |
|---|---|---|
| $HCONH_2$ | m.p. 2.55°C | $d_4^{25}$ 1.12918 |
| M. W. 45.041 | b.p. 210.5°C (decompses) | dielectric const. (25°C) 110.0 |
| | b.p. 104°C/10 torr | |

T.L.V. 20 ppm.; teratogen.

Completely miscible with water. Readily hydrolysed by acids or bases. Extremely hydrogscopic.

*Purification:* (a) Pre-dry by standing for a few hours at room temperature over anhydrous sodium sulphate (200 g $\ell^{-1}$), then distil under reduced pressure (water-pump). (b) Crystallise (by freezing-out) in the absence of moisture and carbon dioxide.

A combination of methods (a) and (b) is particularly efficacious [1].

**Glycerine** (1,2,3-trihydroxypropane, propane-1,2,3-triol)

| | | |
|---|---|---|
| $HOCH_2CH(OH)CH_2OH$ | m.p. 18.18°C | $d_4^{20}$ 1.26134 |
| M. W. 92.095 | b.p. 290.0°C | dielectric const. (25°C) 42.5 |

T.L.V. not known; not very volatile, probably relatively harmless (*cf.* 'ethanol'), unlike ethylene glycol (*q.v.*). No water azeotrope.

Completely miscible with water. Extremely hygroscopic and will take up 50% (by weight) of water from the air.

*Purification:* High vacuum distillation.

**Hexamethylphosphortriamide** (HMPA, hexamethylphosphoramide).

| | | |
|---|---|---|
| $[(CH_3)_2N]_3PO$ | m.p. 7.20°C | $d_4^{20}$ 1.027 |
| M. W. 179.204 | b.p. 233°C | dielectric const. (20°C) 30 |

Caution: in animal tests definitely carcinogenic!

Completely miscible with water and many organic solvents; good solvent for salts.

*Purification:* Pre-dry by standing at room temperature for 24 hours over calcium oxide, barium oxide, or sodium. Then decant and distil under high vacuum.

**n–Hexane**

| | | |
|---|---|---|
| $C_6H_{14}$ | m.p. −95.348°C | $d_4^{25}$ 0.65481 |
| M. W. 86.178 | b.p. 68.740°C | dielectric const. (25°C) 1.8799 |

T.L.V. 100 ppm. Water azeotrope: 61.6°C (94.4% hexane).

At 20°C, it is saturated by 0.0111% water; at 25°C, it is 0.00095% soluble in water.

Commercial hexane in a mixture of hydrocarbons containing *n*-hexane together with a variety of other alkanes, and often contaminated by alkenes and less-volatile materials.

*Purification:* Distillation over sodium hydride dispersion (0.5 g $\ell^{-1}$). To remove traces of alkenes pre-treat by repeatedly shaking with concentrated sulphuric acid (5% by volume), and then washing until neutral.

For u.v. spectroscopy: Pre-treat with sulphuric acid, distil over sodium hydride, and then filter through alumina (basic, activity I, 200 g $\ell^{-1}$).

**Methanol** (methyl alcohol)

| | | |
|---|---|---|
| $CH_3OH$ | m.p. −97.68°C | $d_4^{25}$ 0.78664 |
| M. W. 32.042 | b.p. 64.70°C | dielectric const. (25°C) 32.70 |

T.L.V. 200 ppm. Poisonous if taken orally (5 $cm^3$ may prove fatal; it should not be substituted for ethanol for oral use!!). No water azeotrope. Completely miscible with water.

*Purification:* Place dry magnesium turnings (5 g $\ell^{-1}$) in a flask and add pre-dried absolute methanol (50 $cm^3$). Wait until reaction starts (cloudiness, hydrogen evolution, heat of reaction; a trace of iodine may accelerate this step), then add the methanol requiring purification. Heat slowly to boiling, reflux for three hours, then distil.

**Methyl acetate**

| | | |
|---|---|---|
| $CH_3CO_2CH_3$ | m.p. −98.05°C | $d_4^{25}$ 0.9279 |
| M. W. 74.080 | b.p. 56.323°C | dielectric const. (25°C) 6.68 |

T.L.V. 200 ppm.

At 20°C, 24% soluble in water, and it dissolves 8% water itself.

*Purification:* see 'ethyl acetate'

Methyl acetate forms an addition compound with calcium chloride which is thus not a suitable drying agent for it.

**4-Methylpentan-2-one (methyl isobutyl ketone)**

$(CH_3)_2CHCH_2COCH_3$ m.p. −84°C $d_4^{25}$ 0.7961
M. W. 100.162 b.p. 116.5°C dielectric const. (20°C) 15.11

T.L.V. 100 ppm. Water azeotrope: 87.9°C (75.7% ketone)

At 25°C it is saturated by 1.9% water.

*Purification:* Pre-dry over calcium sulphate (*ca.* 20 g per 100 cm$^3$), decant, then fractionally distil.

**Monoglyme** – see '1,2-Dimethoxyethane'.

**Nitrobenzene**

$C_6H_5NO_2$ m.p. 5.76°C $d_4^{25}$ 1.19835
M. W. 123.112 b.p. 210.80°C dielectric const. (25°C) 34.82

T.L.V. 1 ppm. Very readily absorbed through skin.
Water azeotrope: 98.6°C (12% nitrobenzene).

At 20°C it is saturated by 0.24% water. It is very hygroscopic.

*Purification:* Dry over phosphorus pentoxide (50 g $\ell^{-1}$), decant, then vacuum distil (b.p. 85°C/7 torr).

**n-Pentane**

$C_5H_{12}$ m.p. −129.721°C $d_4^{25}$ 0.62139
M. W. 72.151 b.p. 36.074°C dielectric const. (20°C) 1.844

T.L.V. 600 ppm. Water azeotrope 34.6°C (98.6% pentane).

At 24.8°C it is saturated by 0.0120% water.

*Purification:* see 'hexane'.

**Petroleum ether** (light petroleum)

This term is used for hydrocarbon mixtures boiling over a particular temperature range (e.g. 40–60°C. 60–80°C, 80–100°C; it is *not* an ether. It is often possible to use it as a substitute for pentane or hexane. Flammable, flash point (flame induced) – 57°C, ignition temperature (at a hot surface) 288°C.

*Purification:* see 'hexane'.

**Propan-l-ol** (n-propanol, n-propyl alcohol)

$CH_3CH_2CH_2OH$ m.p. −126.2°C $d_4^{25}$ 0.79975
M. W. 60.097 b.p. 97.20°C dielectric const. (25°C) 20.33

T.L.V. 200 ppm. Water azeotrope: 87.65°C (71.7% alcohol).

Completely miscible with water.

*Purification:* see 'ethanol'.

**Propan-2-ol** (*iso*-propanol, *iso*propyl alcohol)

$(CH_3)_2CHOH$ m.p. −88.0°C $d_4^{25}$ 0.78126
M. W. 60.097 b.p. 82.26°C dielectric const. (25°C) 19.92

T.L.V. 400 ppm. Water azeotrope: 80.10°C (88.0% alcohol.)
Completely miscible with water.

*Purification:* Pre-dry over calcium sulphate (200 g $\ell^{-1}$), decant, add sodium wire or thin shavings (8 g $\ell^{-1}$), and heat slowly to boiling. When alkoxide formation is complete, add *iso*propyl benzoate (35 $cm^3$ $\ell^{-1}$), reflux for 3 hours, then distil. This provides propan-2-ol suitable for Meerwein-Ponndorf reduction. Other procedures: see 'ethanol'.

**Pyridine**

$C_5H_5N$ m.p. −41.55°C $d_4^{25}$ 0.97824
M. W. 79.102 b.p. 115.256°C dielectric const. (21°C) 12.4

T.L.V. 5 ppm. Water azeotrope: 93.6°C (58.7% pyridine).
Completely miscible with water. Hygroscopic.

*Purification:* For many purposes it is sufficient to use pyridine that has been stored over potassium hydroxide. Traces of potassium hydroxide may be removed by distillation. Can also be dried over molecular sieve 4 Å.

**Sulpholane**

$C_4H_8SO_2$ m.p. 28.45°C $d_4^{25}$ 1.2568
M. W. 120.171 b.p. 287.3°C (decomposes) dielectric const. (30°C) 43.3

T.L.V. Not known. Possibly harmful, so avoid contact with skin.

*Purification:* Pre-dry at room temperature overnight with potassium hydroxide (500 g $\ell^{-1}$), decant, and distil under high vacuum (use circulating pump and thermostat at 30°C to cool condenser). The pure material is a solid at room temperature.

**Tetrahydrofuran** (THF)

$C_4H_8O$ m.p. −65°C $d_4^{20}$ 0.8892
M. W. 72.108 b.p. 66°C dielectric const. (25°C) 7.58

T.L.V. 200 ppm. Water azeotrope: 63.4°C (93.3% tetrahydrofuran).
Completely miscible with water and most organic solvents.

*Purification* and removal of peroxides: see 'diethyl ether'.

Peroxide formation is very rapid so tetrahydrofuran should always be freshly distilled before use [2]. A more recent purification method involves refluxing purified tetrahydrofuran with benzophenone and sodium or potassium (or with potassium alone – Care!!) under nitrogen, followed by distillation[3] .

**Tetrahydronaphthalene** (tetralin)

| | | |
|---|---|---|
| $C_{10}H_{12}$ | m.p. −35.790°C | $d_4^{25}$ 0.9662 |
| M. W. 132.207 | b.p. 207.57°C | dielectric const. (20°C) 2.773 |

T.L.V. 25 ppm.

*Purification:* Reflux for one hour over sodium (1%), distil, and then filter through alumina (basic, activity I, 100 g $\ell^{-1}$). Store under nitrogen; it forms peroxides.

**Toluene** (methylbenzene)

| | | |
|---|---|---|
| $C_6H_5CH_3$ | m.p. −94.991°C | $d_4^{25}$ 0.86231 |
| M. W. 92.142 | b.p. 110.625°C | dielectric const. (25°C) 2.379 |

T.L.V. 200 ppm. Water azeotrope : 85°C (79.8% toluene).
At 25°C it is saturated by 0.0334% water.

*Purification:* see 'benzene'.

**Xylene** (mixture of isomeric xylenes)

| | | |
|---|---|---|
| $C_6H_4(CH_3)_2$ | m.p. below −45°C | $d_4^{25}$ *ca.* 0.86 |
| M. W. 106.169 | b.p. *ca.* 140°C | dielectric const. (20°C) *ca.* 2.4 |

T.L.V 200 ppm.

*Purification:* Reflux for 1 hour over sodium (1%), then distil (see 'benzene').

## BIBLIOGRAPHY

[1] D. D. Perrin, W. L. F. Armarego & D. R. Perrin, *Purification of Laboratory Chemicals,* Pergamon Press, 2nd ed., 1980.
[2] *Organic Syntheses,* Coll. Vol. V, 976.
[3] *Techniques of Chemistry,* A. Weissberger ed., Vol. 2, 3rd edition, Wiley, 1976.

# Index